Bioinformatics and Human Genome

Bioinformatics and Human Genome

Dr. Archana Mishra

AUTHORSPRESS

under the auspices of

International Institute of Bioinformatic Studies

First Published 2001
by Authorspress

E-35/103, Jawahar Park (Shree Ganesh Complex),
Laxmi Nagar, Delhi-110 092

e-mail: authorspress@yahoo.com
quantum_soul@mantramail.com

Typeset in NewCenturySchlbk by Chitra Computers

Bioinformatics and Human Genome
ISBN 81-7273-072-1

Printed in India at Tarun Offset Press, Delhi.

Preface

We are living in the age of genetic technology. As we enter the new millennium, the pace of change in human genetics continues to accelerate. It started with the discovery of Mendelism and ended up with the final phase of the draft of the Human Genome Map. The end of the Human Genome Project would mark the beginning of the real understanding of the genetics of mankind. The scope of bioinformatics has also expanded beyond its original coverage. In general, bioinformatics, computational biology, and ancillary computer supports taken together cover the whole spectrum of use of computers in biology-related sciences.

As biological beings, human beings share a common heritage. We all began as a union of chromosomes from our parents, 23 pairs from each. On these chromosomes, composed of deoxyribonucleic acid (DNA), there are an estimated 100,000 genes, of which perhaps as many as 70,000 are expressed in the construction of the brain and nervous system. Thus, any fundamental understanding of our behaviour must consider the contribution that genetics has to offer.

Genetic technology comes into existence as part of the large humanitarian project to cure disease, prolong life, and alleviate suffering. It occupies the moral high ground of compassionate healing. It has aroused considerable public concern. Even people duly impressed by the astonishing achievements of the last decades are nonetheless ambivalent about these new developments. Genetic engineering will deliberately make changes that are transmissible into succeeding generations and may even alter in advance specific future individuals through direct germline or embryo interventions. Genetic engineering may be able, through so-called genetic enhancement, to create new human capacities and hence new norms of health and fitness.

Genomics is generating large quantities of new, high-quality information about complex biological organisms. In addition, computational biology, which uses simulation and computational techniques to define and analyse biological events at the level of the whole organism and, increasingly, at the molecular level, is becoming a central part of biomedical science. To assist their research and help with analyses of the results, biotechnologists rely increasingly on sophisticated IT systems with a range of capabilities.

The Human Genome Project is well underway and, even before its completion, the biotechnology industry is thriving and numerous genetic technology projects are being pursued. Those who have concerns about this rapidly growing new field, are frequently charged with being scientifically ignorant or with holding to antiquated religious and moral ideas. But a close examination of the field does raise serious questions and considerations. Germline gene therapy could change the genetic code of the individual in perpetuity. The technology has the potential to correct genetic flaws at the sperm-egg level but an error in judgement and application would have far reaching influences. More worrisome are plans to introduce desirable traits at the germ level. These facilities will be able available only to the rich and powerful sections of the society and continuous genetic enhancement over generations might led the emergence of a superior race.

Genomic research also has caused a lot of job-related anxieties in many foreign countries. Many people who have been diagnosed to have a genetic diseases might lose their jobs and may not get another replacement. They would not get insurance protection either. Even cell is hair trimming swept out of barber shops would be adequate to provide genome maps. Maps so obtained could reveal the genetic state of the concerned individual and in some cases could be a powerful weapon to threaten people in high positions.

This book guides students and researchers through the very latest developments in this most rapidly changing area of genetics. It contains a veritable mine of new data and features to aid understanding of the principles of human molecular genetics. It will be an excellent companion to all who wishing to gain a background knowledge of the field.

Finally, I acknowledge my indebtedness to my husband Dr. R. C. Mishra, my most exacting critic, for his many valuable suggestions and inspirations to complete the book. I am also much indebted to my publisher K'cherry Sudarshan of Authorspress who only could make this possible in such a short-time. Last but not the least, I wish to express my heartful thanks to International Institute of Bioinformatics Studies (IIBS) for providing data and valuable guidance for the book.

Archana Mishra

Contents

1

Genetics in Human Society

The beginning of science of genetics was made in 1900 by rediscovery of Mendel's laws. Genetics is a science which deals with heredity and variations. It was W. Bateson who used the term 'Genetics' for the first time in 1905. Genetics has played a significant role in shaping human society both historically and in modern times. Modern application of genetics has been particularly important in medicine, agriculture, industry, and forensics. There are associated economic and legal concerns that also affect us all. Genetic analysis has become an integral part of nearly every field of the biological sciences.

Humans recognised long ago that all forms of life bear offspring 'after their kind'. Offspring tend to resemble their parents, usually having a mixture of traits from both of them. It is therefore reasonable to expect that a vigorous and productive plant or animal would produce vigorous and productive offspring. Plant and animal breeding are probably the earliest applications of scientific principles, and are among the most important. People who cultivated their food saved the best individuals for producing future generations.

The science of genetics emerged at the start of the twentieth century. It was Gregor Mendel who first explained the foundation principles of genetics in 1865, based on his meticulous experiments with pea plants. However, no one grasped the importance of his work at the time, and genetics remained in obscurity until 1900, when three botanists independently rediscovered Mendel's principles. In 1905,

William Bateson coined the term genetics to describe the study of inherited characteristics.

During the early part of the 20th century itself, scientists began to apply experimental genetics in a concerted scientific effort to improve domesticated plant and animal species. Astounding gains in agricultural production resulted. While improved cultural practices such as disease control, improved feeding regimes, irrigation, pest and weed control, and fertilisation contributed significantly to these increases, the largest single contributor to improved agricultural production is genetic improvement (brooding).

Genetic improvement of wheat, rice, and a number of other important food crops in less developed countries, coupled with implementation of modern cultural practices, resulted in doubled, tripled, or even quadrupled crop yields over a period of only twenty years. So rapid and influential were these production gains that this international effort became known as the green revolution. But with these gains in agricultural productivity have come social, economic, and environmental concerns, some of which are being addressed by re-evaluating priorities in agricultural breeding and biotechnology programmes.

Scientists have engineered genes that produce human insulin in bacteria for treatment of people afflicted with diabetes. Before genetically engineered human insulin was developed, diabetic patients received insulin extracted from the pancreatic glands of cattle and pigs. However, some patients developed allergies to non-human insulins, which differ slightly from human insulin. Although the genes that encode genetically engineered human insulin were constructed by scientists, the insulin is identical to natural human insulin and does not cause allergic reactions.

Newborn babies are routinely screened for genetic disorders for which early medical intervention can prevent permanent damage. Certain genetic disorders can be identified in human foetuses before birth. Our understanding of the genetic basis of cancer has grown tremendously in the past few years, allowing for development of methods that permit earlier and more accurate detection of cancer. Early detection of cancer makes it possible to begin treatments during the initial stages of tumour development when treatments are most effective.

The Human Genome Project is a large international project aimed at determining the entire human DNA sequence and studying the chromosomal locations and DNA sequence variations for all genes in humans. Given that the human DNA sequence consists of approximately three billion nucleotide subunits of DNA, and that many portions of the DNA must be resequenced using DNA from different individuals to study the extent of sequence variation, this project is no small task. None-the-less, technological improvements and coordination of efforts are contributing to rapid accumulation of sequence data.

Already some genes, such as the one responsible for cystic fibrosis, have been identified and their DNA sequence determined. Such information allows scientists to develop effective treatments for genetic disorders. Many other important genes are now being analysed as the project progresses. Most opponents of the Human Genome Project do not dispute its potential benefits. Instead, they question whether the enormous expense required to complete the project is justified when many other important and beneficial projects could be supported with the funds appropriated to finance it.

Genetics has entered the world of industry. Enzymes often have industrial applications. Enzymes that degrade proteins are added to detergents, and some of these enzymes have been genetically improved to be more effective as detergent additives. Genetically altered bacteria and fungi are being used for waste treatment and hazardous-waste cleanup. Mining companies use plants that have been genetically selected to grow on poor soils to revegetate mine spoils that might otherwise lay barren. Even such items as pigments for cosmetics are produced in genetically selected plant cell cultures.

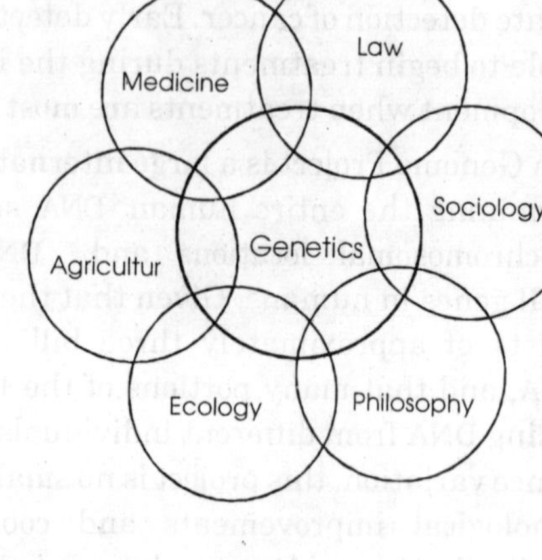

Figure 1.1. Impact of genetics on different areas of human life

The legal issues concerned with genetics are increasing in these days. DNA fingerprinting permits law enforcement officials to identify criminals from small amounts of DNA obtained at crime scenes. Not only can DNA fingerprinting assist in identifying a criminal, it can also exonerate innocent suspects. DNA fingerprinting is also used in paternity testing and in reuniting lost or kidnapped children with their true parents.

With the enormous financial benefits of genetic research, courtrooms are filled with disputes over the ownership of genetically altered organisms and even of the genes themselves. Patents have been granted to protect a company's ownership of genetically engineered bacteria, genetically improved plants and animals, and genes that have been constructed in the laboratory or transferred from one species to another. Even some laboratory techniques have been patented, legally restricting their use and providing royalties to companies that own the rights.

Techniques of genetic analysis have revolutionised several areas of scholarly research extending beyond the boundaries of traditional biology. DNA analysis is now a routine part of some archeological and historical research. Long before there was DNA analysis, human pedigrees derived from information in genealogical libraries helped us understand more about our history, and allowed geneticists to make significant progress in understanding certain inherited traits in humans.

As the molecular foundations of genetics became clear, the relationship of genetics to other major fields of biology likewise clarified. Many of the fundamental principles of cell biology were discovered through genetic analysis of mutations. Cancer is caused by mutations in genes that regulate the cell cycle. Mammals have immunity because certain genes rearrange themselves to produce antibodies and other proteins for the immune system.

Taxonomy has traditionally been based on analysis of structural similarities and differences; these analyses are now augmented by comparisons of similarities and differences in DNA sequence. Ecology and evolutionary biology rely on principles of population genetics, and molecular analysis has become an important tool in studying ecological and evolutionary mechanisms.

REFERENCES

Cook-Deegan, R. M., *The Gene Wars: Science, Politics, and the Human Genome*, New York: Norton, 1994.

Moore, J. A., *Science as a Way of Knowing: The Foundations of Modern Biology*, Cambridge, Mass: Harvard University Press, 1993.

Olby, R. C., *Origins of Mendelism*, 2nd ed. Chicago: University of Chicago Press, 1985.

Orel, V., *Gregor Mendel: The First Geneticist*. Oxford: Oxford University Press, 1996.

Sturtevant, A. H., *A History of Genetics*. New York: Harper & Row, 1965.

2

The Concept of Gene

Genetics has played a significant role in shaping human society both historically and in modern times. The concept of the gene is central to the study of genetics. It is possible to describe a gene as a segment of DNA, a place on a chromosome, something that is passed from parent to offspring, and something that is present among individuals in a population. Although these various descriptions of a gene seem quite different, in each case we are actually referring to the same thing, but at different levels.

Biologists of the nineteenth century viewed the substance of heredity as something fluid, such as blood, semen, or some substance within animal blood or plant sap. They envisioned this substance as something that could be blended and reblended with each generation. Many nineteenth-century biologists also viewed the substance of heredity as changeable rather than constant, and under the influence of the environment. Gregor Mendel was the first to present solid evidence of the constant inherited units that we now call genes.

In Mendel's day, the word gene had yet to be invented. Mendel used the German words *Merkmal* (characteristic, mark, or sign), *Anlage* (plan, design, or predisposition), and *Element* (element) to identify what we now call a gene. He understood genes to be individual hereditary factors that carry the potential for the outward appearance, or phenotype, of an organism. He concluded that genes remain constant from one generation to the next without being changed by the environment. He also concluded that genes are shuffled during formation of the germ cells and at fertilisation, permitting all possible combinations of genes in predictable ratios among progeny, a concept that explains genetic variation among individuals.

When Wilhelm Johannsen coined the word *gene* in 1909, the concept of genes as constant hereditary units that adhere to Mendel's principles of inheritance was already well established. In Mendel's interpretation of his experiments, a single gene corresponds to variation for a single trait. For example, a single gene determines whether the flowers on a pea plant are purple or white. A second gene determines whether a plant is tall or dwarf. This type of phenotypic variation in which contrasting characters fall into discrete classes is known as discontinuous variation. However, among all species there are many traits that do not fall into discrete phenotypic classes. Instead, individuals differ from each other over a continuous range without any discrete phenotypes. This type of phenotypic variation is called continuous variation.

After Mendel's principles were rediscovered in 1900, some biologists assumed that, although discontinuous variation was due to individual genes, continuous variation was due to some form of blending inheritance. William Bateson rejected this explanation and proposed in 1902 that continuous variation derives from the influence of multiple genes on the same trait, and that each of these genes is inherited according to Mendel's principles, a phenomenon called polygenic inheritance.

A few years later, several researchers showed that continuous variation is indeed due to polygenic inheritance combined with the effect of environmental variation. Shortly after Mendel's principles were rediscovered in 1900, Archibald Garrod recognised that the absence of a functional enzyme causes certain inherited disorders in humans. An enzyme is a protein that catalyses a specific biochemical reaction in the cell.

By 1911, Bateson recognised the link between genes and enzymes. He proposed that enzymes themselves are not inherited, but genes are, and somehow a gene determines whether or not a functional enzyme is present in an individual. When a functional enzyme is present, a particular phenotype appears, but when the enzyme is absent, a different phenotype appears.

George Beadle and Edward Tatum further demonstrated the relationship between genes and enzymes in the 1940s and hypothesised that each gene corresponds to an enzyme. How a gene could produce an enzyme, however, remained a mystery. The constancy of genes was called into question shortly after the rediscovery of Mendel's principles. Bateson, at one point, assumed that genes could not be changed but could be lost, resulting in the loss of the enzyme produced by the gene. However, studies with maize and fruit flies revealed that new forms of existing genes occasionally arise by changes called mutations. Constancy of genes appeared to be the rule, but it became evident that genes can mutate and that mutations can be inherited, which creates new genetic variation.

Walter Sutton and Theodor Boveri in 1902 proposed that genes are located on chromosomes and that each chromosome contains many genes. Supporting this proposal, Thomas Hunt Morgan showed in 1910 that a gene responsible for eye colour in *Drosophila melanogaster* is located on the X chromosome. The following year, Alfred Sturtevant, who was an undergraduate studying under Morgan, determined the relative positions of several D. *melanogaster genes on* the X chromosome, beginning the practice of gene mapping. Within a few years, Sturtevant, Morgan, and Calvin Bridges, along with their associates, had mapped numerous genes to their chromosomal locations in *D. melanogaster*.

As the concept of the gene as part of a chromosome became accepted, there was recognition that chromosomes in the nucleus of eukaryotic cells are not the exclusive location of genes. During the first few decades of the twentieth century, researchers found that a few inherited traits in eukaryotes were controlled by genes in the cell cytoplasm rather than genes in the nucleus. Later, these cytoplasmic genes were found to be located on circular DNA molecules in mitochondria or chloroplasts. And other locations for genes were discovered. Prokaryotes (bacteria) lack nuclei and do not have linear chromosomes like eukaryotes (organisms with a cell nucleus). Prokaryotic genes are found in a circular DNA molecule within the cell. Genes were also found in prokaryotic and eukaryotic viruses.

For the first half of the twentieth century, most scientists hypothesised that protein was the genetic material. The first clues about the actual chemical composition of genes surfaced in 1926 when Frederick Griffith discovered that a hereditary substance could be transferred from dead bacteria into living bacteria.

In 1944, Oswald Avery, Colin MacLeod, and Maclyn McCarty demonstrated that Griffith's transforming principle is DNA. In 1952, Alfred Hershey and Martha Chase showed that the hereditary material of bacterial viruses is also DNA. James Watson and Francis Crick found themselves in a fortunate situation in the early 1950s. They suspected that DNA is the genetic material and sought to determine its chemical structure. They had access to unpublished photographs taken by Rosalind Franklin and Maurice Wilkins of X-ray diffraction patterns of DNA that revealed much about its structure.

Watson and Crick used information about DNA chemistry, X-ray diffraction images, and data on DNA composition published by Erwin Chargaff to deduce the chemical structure of DNA in 1953. According to Watson and Crick's model, a single DNA molecule consists of two linear chains of nucleotide subunits with the chains wound around each other to form a double helix. The structure of DNA immediately suggested how DNA could be replicated and soon led to the discovery of what a gene is at its most basic level.

Although proteins are not the genetic material, they are the substance of enzymes and are, therefore, related to genes. Like DNA, proteins are chainlike molecules composed of subunits, but the subunits of proteins are amino acids rather than nucleotides. In the early 1960s, Charles Yanofsky showed that there is a linear relationship between mutations in the nucleotide sequence of a gene and changes in the amino acid sequence of the protein encoded by the gene. These results suggested that genes and proteins are colinear, which means that the linear sequence of nucleotides in DNA corresponds to the linear sequence of amino acids in proteins.

In eukaryotic cells proteins are not synthesised directly from genes in DNA because DNA remains in the cell nucleus while proteins are synthesised on ribosomes in the cytoplasm. There had to be an intermediate between genes and protein. The intermediate is RNA, a molecule very similar in structure to DNA. The information in a gene is copied, through a process called transcription, into an RNA molecule. Then the RNA moves to the cytoplasm, where its information is transferred, through a process called translation, into the linear chain of amino acids constituting a protein. Transcription and translation of genes became collectively known as gene expression.

Biochemical studies demonstrated that each amino acid in a protein corresponds to three nucleotides in a gene. During the late 1950s and early 1960s, a group of scientists, including Marshall Nirenberg, Heinrich Matthaei, Philip Leder, and HarGobind Khorana, deciphered the corresponding triplet nucleotide codes for each of the twenty amino acids. These relationships turned out to be the same in all organisms and are now known collectively as the genetic code. Subsequent research has revealed detailed information about genes at the molecular level.

GENETIC ANALYSIS

Genetics is an analytical science, and the process of discovery unfolds before us when we examine how genetic experiments are conducted and their results interpreted. The experiments of

Gregor Mendel, performed in the 1850s, and his interpretation of his experimental results are a superb example of how genetic analysis can be used to discover and confirm underlying principles.

Mendel's approach of applying designed experimentation with mathematical analysis to answer biological questions was unique in his day. Perhaps for that reason, no one recognised the importance of his experiments during his lifetime. However, once Mendel's work was rediscovered in 1900, his approach to experimentation and analysis set the pattern for much of the genetic research of the twentieth century.

Mendel was a devout clergyman himself, but took a keen interest in evolution. He read much of the scientific literature about plant evolution, speciation, and hybridisation. From his reading he concluded that while many experiments on plant hybridisation had been conducted, none of them were designed in such a way that they could fully answer the question of how hybridisation could produce new species.

Mendel took it upon himself to conduct a large, meticulous, eight-year study of pea plants to address this question. One of his experiments, in which he used tall plants and dwarf plants, serves to illustrate the logic of his design, analysis, and interpretation. Mendel hybridised tall pea plants and dwarf pea plants. The hybrid offspring from these crosses were all tall, a result that didn't surprise Mendel, for many of his predecessors had seen similar results. He then allowed these hybrid plants to self-fertilise. In the second-generation progeny, he counted 787 tall plants and 277 dwarf plants, approximating very closely a ratio of three tall plants to one dwarf plant.

Mendel performed similar experiments with six other traits, and each time the same pattern appeared: all of the first-generation hybrid progeny had only one of the two contrasting characters, but the other character reappeared in the second generation in approximately one-quarter of the individuals.

Mendel concluded that each plant carried hereditary factors that influenced the traits he was studying. When plants with contrasting characters were crossed, each parent plant contributed a factor to the offspring so that the hybrid offspring carried both factors simultaneously. He concluded that both factors were present in the hybrid, and that one of these factors was dominant and the other recessive. Rather than speaking of factors, we now say that a hybrid for plant height has a dominant allele and a recessive allele of the gene controlling plant height. Different alleles are defined as different versions of the same gene.

Mendel represented the dominant allele with an uppercase A and the recessive allele with a lowercase a. He then represented the first-generation hybrid plants as Aa because they carried both alleles. Only the character associated with the dominant allele appeared in the phenotype of the hybrid plants. In the plant height example, the hybrid plants simultaneously carried the allele for tall stature and the allele for dwarf stature, but the tall-stature allele was dominant, so all the first-generation progeny were tall.

Mendel developed his interpretation in light of the cell theory of his day. He conclude that during formation of the germ cells in the hybrid plant, the dominant and recessive alleles segregated from each other so that half of the pollen grains received the dominant allele A and the other half received the recessive allele a. The same was true for the egg cells; half received A and half a. Under these conditions, random union of pollen cells and egg cells at fertilisation should result in a mathematical series of all possible combinations of alleles in the second-generation progeny. Mendel represented this series as:

$$\frac{A}{A} + \frac{A}{a} + \frac{a}{A} + \frac{a}{a}$$

Here the symbols in the numerator of each fraction denote the alleles from the pollen cells and the symbols in the denominator of each fraction denote the alleles from the egg cells. With random union of pollen and egg cells at fertilisation, each of these four possible combinations has an equal probability of appearing in the second-generation offspring of a hybrid plant. Three of the four combinations above have at least one dominant A allele, which causes tall plants. One of the four combinations (a/a) has only a, which causes dwarf plants. Under this theory, about three-fourths of the progeny should be tall, while the remaining one-fourth should be dwarf. This explained the 3:1 ratio that Mendel observed in the second-generation progeny for each of the seven traits he studied.

Having developed a theoretical interpretation to explain his results, Mendel designed additional experiments to test his interpretation. The results of these additional experiments supported his interpretation, lending it additional credibility. He also discovered through his experiments that the inheritance of each set of contrasting characters was independent of the inheritance of the other sets of contrasting characters. For example, the inheritance of plant height was independent of the inheritance of flower colour. He interpreted this to mean that all possible combinations of characters could be obtained in the second-generation offspring from hybrids.

Mendel concluded that hereditary factors in hybrids segregated into germ cells in a 1:1 ratio and that the random union of these cells resulted in second-generation phenotypes in a 3:1 ratio due to dominance. In deriving this conclusion, he interpreted his results in light of the cell theory that was common in the scientific literature of his day Mendel also concluded that inheritance followed predictable mathematical patterns, a concept that was completely novel in his day but was clearly demonstrated by his results.

Mendel made no attempt to explain what the hereditary factors were or to speculate about their composition. He also did not attempt to explain how an underlying factor could cause an observable phenotype, or how the cells could carry and express such factors. This lack of speculation makes sense because his results, and the results of others, provided no information about these phenomena.

Mendel concluded his paper by attempting to reconcile the observations of other plant hybridists in the light of his theory and to explain how hybridisation and the subsequent new combinations of genes could result in new species of plants. Mendel's approach illustrates some of the fundamental aspects of genetic analysis that are as applicable today as they were for him well over a century ago. He chose an appropriate organism for the experiments he had in mind. The seven traits he chose to analyse could be identified easily and unambiguously and were not influenced by environmental variation. This allowed Mendel to focus on heredity alone, without the confounding effects of environmental influences, a point that he stressed in his paper as crucial to the success of his experiments.

He recognised that in order to obtain valid information, he needed to analyse large numbers of individuals. He designed his experiments to do this, and appropriately applied mathematical analysis to his results. He then developed an interpretation based on his results and on the scientific theories of his day, and tested his interpretation by further experimentation. He concluded his report by placing his experimental results and their interpretation in the context of previous experimentation.

THE GENETIC CODE

Proteins are the final products of most genes and constitute the most diverse class of biological molecules. Many proteins act as enzymes, and enzymes regulate nearly every biochemical reaction. Some proteins are responsible for the transport of substances across cell membranes. Other proteins may accumulate where they are stored for later use, such as in bird eggs and plant seeds. Certain proteins act as structural components, such as those that compose the bulk of muscle tissue, hair, and fingernails. Whatever their functions, all proteins are encoded by genes and are synthesised through translation.

Amino acids are the building blocks of protein. They have a common structure consisting of a central carbon atom with a carboxyl group, an amino group, and a hydrogen atom bound to it. This leaves one binding site to which a side group may bind. The side group of an amino acid is denoted generically as the R group and is the only part of the molecule that differs from one amino acid to the next. There are many different amino acids, but only 20 that are utilised in translation. Proteins consist of varying combinations of these 20 amino acids. Each of the amino acids utilised in translation is distinguished by its R group, and the chemical properties of the R groups define categories of amino acids.

In the process of translation, amino acids are attached to each other through the formation of peptide bonds. A peptide bond forms in an enzyme-catalysed reaction between the carboxyl group of one amino acid and the amino group of another. This reaction takes place on the ribosome during translation. The reaction removes a water molecule as it forms the peptide bond, which is a covalent bond between the carbon in the carboxyl group of one amino acid and the nitrogen in the amino group of another amino acid. Many amino acids may attach to each other by peptide bonds, forming a long chain called a polypeptide. Each polypeptide has an amino group on one end and a carboxyl group on the other end. Polypeptides are synthesised in the amino-carboxyl direction, and amino acid sequences are written in this direction.

The 'message' in a messenger RNA is the segment of nucleotides that translates into a protein. This implies that there is a correspondence between the information contained in the mRNA, which is composed of a specific sequence of nucleotides, and the information contained in a polypeptide, which is composed of a specific sequence of amino acids.

In fact, the mRNA and the polypeptide are colinear, which means that the 5'→ 3' linear sequence of nucleotides in a segment of the mRNA corresponds directly to the amino → carboxyl linear sequence of amino acids in the polypeptide. Convinced that the nucleotide sequence in mRNA determines the amino acid sequence of a protein, we may ask, how many nucleotides are needed to specify each amino acid? This was the first question facing researchers in the 1950s who wanted to decipher the genetic code, the pattern by which nucleotides in an mRNA correspond to the amino acids in the encoded protein.

Characteristics of Genetic Code

While proteins are composed of varying combinations of up to 20 different amino acids, RNA is composed of only four nucleotides. If one nucleotide corresponded to one amino acid, it would be possible to code for only four amino acids.

If each amino acid were specified by a combination of two nucleotides, it would be possible to code for 16 amino acids because there are 41, or 16, possible combinations of two nucleotides each. We are still four combinations short of specifying all 20 amino acids. If we increase the number of nucleotides per amino acid to three, then there are 41, 64 or far more possibilities.

The genetic code indeed consists of triplet nucleotide combinations. This means that the genetic code must have some redundancy in it. In other words, some amino acids may each be encoded by several triplet combinations of nucleotides.Due to this redundancy, the genetic code is said to be a degenerate code. The questions, of which triplets of nucleotides code for which amino acids, were all answered in the early 1960s when the genetic code was deciphered.

The genetic code has 64 triplet nucleotide combinations, each called a codon. All but three of the 64 codons specify amino acids. The three codons that do not specify amino acids (UAA, UAG and UGA) are signals for termination of translation and are therefore called termination codons (or stop codons). Figure 2. 1 gives the genetic code.

One of the most intriguing aspects of the genetic code is its near universality. The same genetic code is used for translation in every organism from bacteria to mammals, with only a few isolated exceptions. As the genetic code is preserved in all forms of life, it has strong evolutionary significance. Because the genetic code of humans is identical to the genetic code of bacteria, it is possible to synthesise large quantities of valuable proteins from human genes that have been inserted into bacteria.

second base of codon

first base of codon	U		C		A		G		third base of codon
U	UUU	Phe F	UCU	Ser S	UAU	Tyr Y	UGU	Cys C	U
	UUC		UCC		UAC		UGC		C
	UUA	Leu L	UGA		UAA	Ter	UGA	Ter	A
	UUG		UCG		UAG		UGG	Trp W	G
C	CUU	Leu L	CCU	Pro P	CAU	His H	CGU	Arg R	U
	CUC		CCC		CAC		CGC		C
	CUA		CCA		CAA	Gln G	CGA		A
	CUG		CCG		CAG		CGG		G
A	AUU	lle I	ACU	Thr T	AAU	Asn N	AGU	Ser S	U
	AUC		ACC		AAC		AGC		C
	AUA		ACA		AAA	Lys K	AGA	Arg R	A
	AUG	Met M	ACG		AAG		AGG		G
G	GUU	Val V	GCU	Ala A	GAU	Asp D	GGU	Gly G	U
	GUC		GCC		GAC		GGC		C
	GUA		GCA		GAA	Glu E	GGA		A
	GUG		GCG		GAG		GGG		G

Figure 2. 1. The genetic code: the three-letter and one-letter abbreviations of the amino acids. The initiation codon, AUG, is shaded darkly and the three termination codons—UAA, UAG, and UGA—are shaded lightly.

Reading Frame for Translation

The genetic code is non-overlapping. Once the reading frame is established from the initiation codon, it proceeds every three nucleotides from that point on, adding amino acids unit a termination codon is reached. To translate an mRNA, ribosomes bind to the mRNA and read the codons that specify the polypeptide.

Translation does not begin immediately at the 5' end of the mRNA. Instead, somewhere near the 5' end is an initiation codon (or start codon) that codes for the first amino acid in the polypeptide. The initiation codon is usually the codon AUG, which codes for methionine in eukaryotes and n-formyl methionine in prokaryotes. At sites beyond the initiation site, the codon AUG codes for methionine. The initiation codon not only indicates where translation begins, it also establishes the reading frame for translation.

The reading frame is governed by two facts:

i) codons in the mRNA do not overlap, and

ii) there are no intervening nucleotides between codons.

In other words, the next three nucleotides after the initiation codon form the second codon, and the following three nucleotides form the third codon, and so on-a pattern that continues uninterrupted throughout the reading frame.

For instance, suppose that the following sequence of ribonucleotides, which includes the initiation codon, is part of an mRNA:

5'. . . GCCACCAUGGCUGGAGUCAC. . . 3'

Bolface AUG sequence is the initiation codon and codes for methionine. The second codon is composed of the next three nucleotides, GCU, and codes for alanine. The third codon, GGG, codes for glycine. Translation continues moving in the 5' → 3' direction, adding amino acids according to each triplet codon to produce:

5'... GCCACC**AUG**GCUGGGAGUCAC ... 3'

methionine-alanine-glycine-serine-histidine ...

At the first two codons, AUGGCU, we can find the sequence GGC within it (AUGGCU). If these three nucleotides were translated as a codon, they would code for glycine. However, as codons do not overlap, the nucleotide sequence GGC is out of frame and will not be recognised as a codon. Translation continues from codon to codon uninterrupted until a termination codon is reached. At this point, the ribosomes separate from the mRNA, and transcription ceases. Since termination codons do not code for any amino acid, the amino acid encoded by the codon just before the termination codon is the last amino acid in the polypeptide.

Role of Ribosomes in Translation

Ribosomes are the sites of translation. They are composed of rRNA molecules and numerous proteins. The actual number of rRNAs and proteins and the final size of the intact ribosome vary among species. Although there are many more proteins than rRNAs in a ribosome, the rRNAs are much larger than the proteins and generally account for over 60 per cent of the mass of the ribosome. Prior to translation, the rRNAs and proteins are assembled into two separate subunits, the large subunit and the small subunit. These subunits form an intact ribosome by attaching to an mRNA during initiation of translation.

Both prokaryotic and eukaryotic ribosomes have large and small subunits with similar structures, although ribosomes are somewhat larger in eukaryotes than in prokaryotes. The *E. coli* ribosome has been well studied and is representative of prokaryotic ribosomes. The intact ribosome has a Svedberg value of 70S and consists of a 50S large subunit and a 30S small subunit. The large subunit contains two rRNAs, a 23S rRNA (2904 nucleoticles) and a 5S rRNA (120 nucleotides), along with 31 proteins. The small subunit contains a single 16S rRNA (1541 nucleoticles) and 21 proteins. Each subunit has a specific three-dimensional shape that allows the two subunits to interlock with each other.

Mammalian ribosome is the best characterised of eukaryotic ribosomes. The intact ribosome has a Svedberg value of 80S with a large 60S subunit and a small 40S subunit. The large subunit contains three rRNAs, a 28S rRNA (4, 718 nucleotides), a 5.8S rRNA (160 nucleotides), and a 5S rRNA (120 nucleotides). The small subunit contains an 18S rRNA (1, 874 nucleotides). The number of proteins in eukaryotic ribosomes varies among species, but is about 45-50 in the large subunit and 30-35 in the small subunit.

In the process of translation, several ribosomes may attach one after the other onto an mRNA and proceed through translation as a chain of ribosomes, each ribosome translating one polypeptide. When this happens, the assembled ribosomes and the mRNA together are called a polysome.

THE T-RNA

Ribosome itself cannot place the appropriate amino acids in the growing polypeptide chain. This is the function of transfer RNAs (tRNAs). Each amino acid attaches to a tRNA that is responsible for carrying the amino acid to the ribosome and placing it in the proper order according to the nucleotide sequence in the mRNA.

t-RNAs are quite small, only 74-95 nucleotides, and they all have the same general structure. Following transcription and processing, over half of the nucleotides base-pair with one another to form a cloverleaf shape when drawn in two dimensions. The cloverleaf consists of four major arms and a small extra arm. They are:

i) acceptor arm,

ii) anti-codon arm,

iii) D arm and

iv) T ψ C arm.

Acceptor arm: It is so named because it forms the part of the molecule that accepts the amino acid. It includes both the 5' and 3' ends of the linear molecule and has four impaired nucleotides on the 3' end extending beyond the paired nucleotides. The final three nucleotides on the 3' end are CCA. The amino acid attaches to the terminal A on the 2' or 3' hydroxyl group. This position on the acceptor arm where the amino acid attaches is called the amino acid attachment site.

Anti-codon arm: The anti-codon arm lies opposite the acceptor arm in the cloverleaf and consists of a base-paired stem and a loop of seven impaired nucleotides, the centre three of which form the anti-codon. This site is critical because the three nucleotides in an anti-codon pair with the three nucleotides in a codon in the mRNA during translation.

D arm and the TψC arm: Each of these arms have a base-paired stem and a loop of unpaired nucleotides. These arms help form the three-dimensional structure necessary for enzymes to recognise the tRNA and for the tRNA to interact with the ribosome. The Greek letter ψ (psi) refers to the modified base pseudouracil that is between a T and C in the impaired loop of the TψC arm. The T is thymine that is initially transcribed as U but later methylated to form T. The letter D refers to dihydrouracil, a modified nucleotide in the D arm. Another small extra arm is also present there, usually consisting entirely of unpaired nucleotides, between the anti-codon arm and the TψC arm.

The cloverleaf shape of tRNAs exists only in two-dimensional representations of their base pairing. In the cell, tRNAs assume a three-dimensional structure that has an L shape with the amino acid attachment site at one end and the anti-codon at the other. The TψC arm and the D arm are adjacent to one another on the outside corner of the L. The three-dimensional structure of all tRNAs must be essentially the same, because they all must fit into the same sites in the ribosome during translation.

However, there are minor structural differences that distinguish one type of tRNA from another. Represented in two dimensions, transfer RNAs have a cloverleaf shape with an amino acid attachment site in the acceptor arm and an anti-codon in the anti-codon arm. The three nucleotides in an anti-codon pair with the three nucleotides in a codon of the mRNA during translation.

One particular tRNA accepts only one of the 20 amino acids at the amino acid attachment site—the amino acid that corresponds, according to the genetic code, to the sequence of nucleotides in the anti-codon. For example, a tRNA with the anti-codon sequence 3' UAC 5' accepts only methionine, because its anti-codon pairs with AUG, the codon for methionine.

A tRNA may be free, with no amino acid attached, or it may have its specific amino acid bound to the amino acid attachment site on its acceptor arm. A charged (aminoacylated) tRNA is one that has its amino acid attached. A charged tRNA has its amino acid attached by a bond between the 2' or 3' hydroxyl group on the 3' end of the tRNA and the carboxyl group of the amino acid. Charging is catalysed by enzymes called aminoacyl-tRNA synthetases. These enzymes bind amino acids to the amino acid attachment site through a two-step process.

First, the aminoacyl-tRNA synthetase causes the amino acid to react with ATP to form aminoacyl adenylate, which is an amino acid with an adenosine nucleotide attached to its carboxyl group by a phosphodiester bond. Energy for the reaction is derived from breaking the phosphate bonds in the ATP molecule that contributes the adenosine. At this point the amino acid is activated, or in other words, prepared for attachment to the tRNA. Second, the synthetase bonds the activated amino acid to the amino acid attachment site on the tRNA, releasing adenosine monophosphate (AMP) in the process.

The aminoacyl-tRNA synthetase specific for an amino acid also recognises all the different tRNAs that carry that amino acid. Certain amino acids encoded by more than one codon may be carried by tRNAs with different anti-codons. For example, the amino acid serine is encoded by six different codons and has three different tRNAs to match these codons. Each of these tRNAs is recognised by the aminoacyl-tRNA synthetase that is specific for serine. Different tRNAs that carry the same amino acid are called cognate tRNAs.

Degeneracy of t-RNA

The genetic code is said to be degenerate. Since there are 61 codons that specify amino acids, there might be 61 anti-codons and, therefore, 61 different tRNAs. However, the actual number of anti-codons and tRNAs is much lower because codon-anti-codon pairing in the ribosome is precise only for the first two nucleotides of the codon. Codon-anti-codon pairing at the third position in the codon does not follow the pairing rules as closely as pairing at the other two positions. Within the ribosome, the conformation of the anti-codon loop is such that some anti-codons can pair with two, or even three, codons. The base-pairing rules that apply only to the third nucleotide in codon-anti-codon pairing are provided in Table 2. 1.

The set of base-pairing rules in Table 2. 1. is called the wobble hypothesis. There are several interesting features of the wobble hypothesis. The two purines, A and G, in the codon may both pair with U in the wobble site of the anti-codon. The two pyrimidines, U and C, in the codon may both pair with G in the wobble site. Three nucleotides in the codon, U, C, and A, may pair with a modified nucleotide, inosine (I), in the wobble site.

Inosine is a modified purine that is similar to guanine but lacks the amino group attached to the number 2 carbon in guanine.

Table 2. 1.
Codon-anti-codon pairing at the third position in the codon according to the Wobble hypothesis.

5' nucleotide in anti-codon	*3' nucleotide in codon*
U	A or G
C	G
A	U
G	U or C
I	U, C, or A

The wobble hypothesis, which explains the pattern of degeneracy in the genetic code, provides pairing rules for the third nucleotide in a codon with its corresponding nucleotide in the anti-codon.

DECIPHERING THE GENETIC CODE

In the mid 1950s, the logical assumption was that the linear sequence of nucleoticles in DNA corresponds to the linear sequence of amino acids in the protein through some sort of genetic code. Using mathematical logic, geneticists determined that at least three nucleotides were needed to code for one amino acid, at least in some cases. Almost nothing was known about tRNA during the 1950s, so it was generally assumed that amino acids were assembled directly on the mRNA molecule. This required not only a linear but also a physical correspondence between amino acids and the nucleoticles that coded for them. The spacing of amino acids in a polypeptide chain is about the same as the spacing of nucleotides in DNA, implying that a one-to-one correspondence between nucleotides and amino acids was likely.

As three nucleotides were needed for each amino acid, a genetic code of overlapping triplet codons was the only way to reconcile triplet codons with a one-to-one correspondence of nucleotides and amino acids. For example, the RNA sequence AAGACGCUAC would have eight overlapping triplet codons: AAG, AGA, GAC, ACG, CGC, GCU, CUA, and UAC, moving one nucleotide along the molecule for each codon. During 1957, partial amino acid sequences for several proteins had been determined.

Using these sequences, Sydney Brenner compared each aminoacid with its neighbours and calculated the number of overlapping triplet codons that would be necessary to code for the known sequences. The minimum number of triplet codons required was 70, which is more than the 64 possible combinations—meaning that a genetic code based entirely on overlapping triplet codons was not possible.

It was logical to assume that a mutation in the DNA would alter the amino acid sequence for a protein. Under the assumptions of the overlapping codon hypothesis, a single-nucleotide substitution mutation should affect three adjacent amino acids, because the altered nucleotide is part of three overlapping triplet codons. However, if codons do not overlap, a single-nucleotide substitution mutation should affect only one amino acid.

Using some rather ingenious methods, the genetic code itself was deciphered during the early 1960s. Today the nucleotide sequence of a segment of DNA and the amino acid sequence of a polupeptide may be determined routinely. Marshall Nirenberg and Heinrich Matthaei developed the first of these indirect approaches in 1961 by preparing a cell-free protein synthesis system from *E. coli.* They ground the bacterial cells with an abrasive to break them open and release the cellular contents. These cellular contents made a cell-free extract that contained active ribosomes as well as all other necessary components for protein synthesis. They then prepared synthetic RNAs using an enzyme, called poly-nucleotide phosphorylase, that builds an RNA chain out of nucleotides without a DNA template, so that the nucleotides are incorporated at random. By providing the enzyme with only one type of nucleotide, they could produce a synthetic RNA molecule consisting only of that nucleotide.

Research groups made synthetic RNAs consisting of two or three different nucleotides. The degeneracy of the genetic code soon became evident. Synthetic RNAs containing random combinations of U and A, U and C, or U and G all included leucine in the polypepticles they produced, meaning that several combinations of these nucleotides must code for leucine. Later, several tRNAs were isolated and combined with synthetic mRNAs in cell-free systems. Researchers then determined which nucleotides in the mRNA paired with the anti-codon in each tRNA.

In 1963, Nirenberg put the information from these various studies together and identified the amino acids associated with the nucleotide composition of 40 codons. When the actual genetic code was deciphered, it turned out that the nucleotide compositions of 35 of the 40 codons proposed by Nirenberg were correct.

In 1965 Nirenberg and Philip Leder discovered that ribosomes would read a synthetic mini mRNA, consisting of three nucleoticles (one codon) and would hold the appropriately charged tRNA in the ribosome with its anti-codon bound to the synthetic codon. Thus each codon could be tested individually to determine which amino acid it retained in the ribosome. Nirenberg and Leder added each of the 64 possible mini mRNAs to a cell-free system and allowed the ribosomes to bind the tRNA charged with the appropriate amino acid. They then separated the ribosome-mini-mRNA-tRNA-amino acid complexes from the remaining components and determined which amino acid was bound in the ribosomes. Using this ribosome-binding test, they identified unambiguously 50 of the 64 possible triplet codons, and to narrow the possibilities for the remaining codons.

In 1966 HarGobind Khorana and his co-workers performed some particularly interesting experiments using synthetic mRNAs. They constructed synthetic polymicleotides consisting of repeating nucleotide combinations, for example, poly (UC), a nucleotide made of repeating UC units (UCUCUCUCUC... etc.) coded for a repeating polypeptide of alternating serine and leucine (serine-leucine-serine-leucine-... etc.). No matter where translation starts, UCU and CUC codons alternate within the reading frame. Khorana and his co-workers confirmed that the portion of the genetic code determined by the ribosome-binding test.

The genetic has been confirmed in thousands of studies and is nearly universal among all species. Until the mid-1970s, the code was thought to be completely universal. However, comparison of the DNA sequence of several genes with the amino acid sequence of proteins revealed a number of exceptions to the genetic code.

In the prokaryote *Mycoplasma capricolum*, the codon UGA is read as tryptophan instead of a termination codon. In the eukaryotic ciliates (protozoans), UAA and UAG are read as glutamine instead of termination codons. The protein synthesis systems contained in mitochondria of various organisms appear to contain one or more differences from the standard genetic code. These differences are not universal in the mitochondria of all species, however, and appear to be restricted to certain phylogenetic groups such as mammals or yeasts.

At least some of the perceived exceptions to the genetic code are actually due to specific nucleotide modifications in mRNA after it is transcribed, a process called mRNA editing. For nearly all genes, the mRNA sequence matches the sequence in the DNA that encoded it. However, for a few genes, mRNA editing causes a nucleotide or two to differ between the sequence in the mRNA and the corresponding sequence in the DNA.

Once the mRNA is edited, a codon in the mRNA may be different from the corresponding nucleotide sequence in the DNA. When DNA sequence and protein sequence are compared without comparing the mature mRNA produced in the cell, the difference appears to be an exception in the genetic code. In reality, the mRNA may have been altered, rather than the genetic code. In addition, there are true exceptions in the genetic code that cannot be explained by mRNA editing, indicating that the genetic code is not entirely universal.

GENE MUTATIONS

A mutaion is an abrupt qualitative and quantitative change in the genetic material of an organism. The word 'mutation' is derived from the Latin verb *mutare*, which means 'to move or change'. A genetic mutation is a change in the genetic information in DNA. The word mutation was first used in genetics to describe spontaneous heritable changes in the appearance of an organism. A mutation in the DNA of a gene may alter the amino acid sequence encoded by that gene, which may alter or eliminate the function of the gene, which, in turn, may affect the phenotype of an individual who inherits the mutation.

Types of Mutations

Mutation is the basis for evolution and the only means by which new genetic variation is created. It is a heritable change in the nucleotide sequence of a cell's DNA. The change may be as small as a single-nucleotide substitution, such as a G in place of an A or as large as the addition or deletion of millions of nucleotides. Since DNA is the substance that mutates, mutations are faithfully replicated from one cell generation to the next and from one organismal generation to the next. Mutations may be intragenenic (point mutation) or intergenic.

Point mutation

Point mutation is the change in a single nucleotide or in a few adjacent nucleotides in the DNA; for this reason, a point mutation usually affects only one gene. There are many different types of point mutations, and many ways in which point mutations arise. Chromosomal aberrations differ significantly from all other types of mutations. Chromosomal changes eliminate, duplicate, or rearrange chromosomal segments that may include hundreds of thousands of nucleotide pairs and several genes. The effects of chromosomal changes are quite different from those of point mutations.

There are three types of point mutations, substitutions, deletions, and insertions, drawing examples from the human PAH gene, which encodes the enzyme phenylalanine hydroxylase. Mutations in human PAH gene are responsible for the genetic disorder phenylketonuria (PKU), which causes severe mental retardation if left untreated.

Substitution mutations: Substitution mutation is the most common type of point mutation, in which one nucleotide pair is substituted for another. DNA usually mutates during replication, and a nucleotide in only one strand of the DNA mutates. However, once the DNA replicates again, the mutation appears in both strands of one of the progeny molecules and is replicated faithfully in both strands from that point on. Geneticists usually study mutations in DNA that has replicated many times since it mutated, so they observe the mutation as a substitution in both strands when compared to the original double-stranded DNA sequence.

Substitution mutations are usually designated by writing the nucleotide change with an arrow, such as G → A, where G is the original nucleotide and A is the nucleotide substituted for it. Only the sense-strand sequence is written although the mutation appears in both strands. A G → A substitution in the sense strand implies a C → T substitution in the anti-sense strand. The effect that a substitution mutation has on gene function depends on where in the gene the DNA mutates.

A mutation that makes no change in the amino acid sequence is called a same-sense mutation. Same-sense mutations are also silent mutations, mutations that do not alter protein function. A mutation that causes an alteration in the amino acid sequence of a gene's product is called a missense mutation. Most missense mutations alter the function of a gene. However, some missense mutations cause amino acid substitutions that do not significantly alter protein function and are, therefore, silent mutations. A missense mutation that creates a new termination codon is called a nonsense mutation.

Substitution mutations are either transitions or transversions. A transition is a purine → purine substitution (A → G, G → A), or a pyrimidine → pyrimidine substitution (T → C, C → T). A transversion is a purine → pyrimidine substitution (A → T, A → C, G → T, G → C) or a pyrimidine → purine substitution (T → A, T → G, C → A, C → G). Transitions are more frequent than transversions.

Deletion and insertion mutations: In deletion mutations, one or more nucleotides are deleted from the DNA, and insertion mutations, in which one or more nucleotides are inserted into the DNA. Any deletion or insertion that is not a multiple of three nucleotides, causes a frameshift, provided the mutation is located in the reading frame.

Deletions or insertions that are multiples of three nucleotides cause corresponding deletions or additions of amino acids in the polypeptide but do not shift the reading frame. Frameshift mutations take the original termination codon out of frame and bring a new termination codon into frame, shortening or lengthening the polypeptide depending on where the first termination codon appears in the mutant reading frame. Mutations do not have to be in the reading frame to affect the polypeptide. Conserved sequences that lie outside of the reading frame are important for initiation of transcription, intron removal, polyadenylation, and initiation of translation. Mutations in conserved sequences may affect the polypeptide encoded by the gene.

Trinucleotide repeat expansions: Trinucleotide repeat expansion, which denotes an increase in the number of trinucleotide tandem repeats, was discovered in humans in 1991. A genetic disorder called fragile X syndrome was found to be due to expansion of a region in DNA where the trinucleotide sequence CGG is repeated several times in tandem. Over 10 genetic disorders in humans are now known to be caused by trinucleotide repeat expansions.

Fragile X syndrome is the most thoroughly studied and offers a good example of some peculiar effects that arise from this type of mutation. It is caused by disruption of a gene on the X chromosome called FMR1 (for fragile X mental retardation 1), which encodes the protein FMRP. This protein binds to mRNAs in brain cells and apparently affects the expression of a number of genes essential for proper brain development. Within the untranslated region at the 5' end of the gene is a three-nucleotide sequence (CGG) that is repeated anywhere from six to 50 times in most people, with 29 copies being the most common number.

People with a normal number of repeats show no symptoms of the disorder. The number of repeats can increase or decrease slightly from generation to generation, but rarely exceeds 50. This normal fluctuation has no effect on the size of the polypeptide because the repeated region appears in the mRNA upstream from the reading frame. Some people, however, may acquire an abnormally large number of repeats. Repeats ranging from 50 to 230 constitute what is known as a fragile X premutation.

People with the premutation rarely show any symptoms of the disorder, but the descendants of females who carry the premutation have an increased risk of fragile X syndrome. People who have fragile X syndrome often have as many as 700 CGG repeats and sometimes over 1, 000.

When the repeated region expands significantly beyond the premutation size, a fragile X full mutation results. The bases in the nucleotides of the repeated region and the promoter become highly methylated, which inhibits transcription of the FMR1 gene. Ribosomes do not efficiently translate the few mRNAs that are produced. The amount of FMRP produced under these conditions is inadequate and symptoms of fragile X syndrome appear. Not all cases of fragile X syndrome are due to trinucleotide repeat expansion. In one case, a missense substitution mutation that substitutes asparagine for isoleucine is responsible. Several deletions of either the entire gene or large portions of it are also known to cause fragile X syndrome.

There are several other genetic disorders in humans whcih are due to trinucleotide repeat expansion. Huntington disease is characterised by progressive neurological degeneration that usually does not begin until after the age of 30. Once the degeneration begins, the affected person progressively loses motor function, behaves in unusual ways, becomes demented, and eventually dies from the disease, usually within 10-15 years. Huntington disease is usually due to expansion of a CAG repeat in the coding region of the *IT15* gene, which encodes a protein named huntingtin. Each CAG is in frame and is translated as a glutamine. Affected persons have 36-121 repeats, and the lengthened string of glutamines alters the protein's function. Unaffected persons have 10-35 repeats that are translated as a string of glutamines.

Mutations and protein function: The effect of a mutation on protein function depends on several factors: which amino acids in the polypeptide chain are altered, how many amino acids are altered, and what specific changes in amino acid sequence arise from the mutation. A same-sense mutation has no effect on the function of a protein because the amino acid sequence remains unchanged.

Missense mutations, however, can have a variety of effects, ranging from no loss of function to complete loss of function or, more rarely, a gain of function. The final effect depends on how and where the mutation alters the protein. A protein's function depends predominantly on the physical shape of the protein, as determined by its primary, secondary, tertiary, and quaternary structures, the chemical properties of the protein, and the active site of the protein if it is an enzyme.

Missense mutations that do not significantly alter any of these will probably have little, if any, effect on the protein's function. For example, valine and isoleucine are both nonpolar amino acids, differing from one another by a single methyl group. Their chemical properties are nearly identical. A missense mutation that substitutes valine for isoleucine should have little, if any, effect on protein function. However, a mutation that alters the protein's active site, or alters the tertiary or quaternary structure, may change the protein's ability to carry out its function or, in many cases, eliminate the protein's function altogether.

Most frameshift mutations completely eliminate protein function because they alter nearly every amino acid from the point of the mutation onward and usually change the location of the termination codon. Even if a frameshift mutation does not affect the active site of an enzyme, it usually alters the structure of the enzyme drastically, causing the enzyme to lose its function.

Mutations that have an effect on a protein's activity are loss-of-function mutations, which either reduce the protein's ability to function or eliminate its function altogether. Most of the hundreds of known genetic disorders in humans are caused by loss-of-function mutations. More rarely, a mutation may cause a gene to gain function compared to its previous state, usually by producing a greater-than-normal quantity of protein. These are called gain-of-function mutations.

Mutations outside transcribed region of gene: Mutations outside of the transcribed region, as a general rule, do not alter the amino acid sequence of a protein. However, depending on where a mutation is located, it may affect the amount of the protein that is produced, thus affecting its level of activity in the cell. For example, a mutation within the promoter region may prevent RNA polymerase from binding and thus prevent transcription, eliminating the protein's synthesis entirely. A promoter mutation may also reduce or, more rarely, enhance the ability of RNA polymerase to transcribe the gene, thus affecting how much protein is produced. In fact, most gain-of-function mutations reside in the promoter or upstream regulatory regions of a gene.

Forward mutations: A forward mutation changes an original, usually functional, form of a gene to a mutant form, in which the DNA sequence differs from the most common sequence. Forward mutations are usually detected as a loss of protein function, often due to a change in the protein's amino acid sequence.

Reversions: A second mutation in the mutant form of a gene may cause the gene to revert back to its functional form, in which it encodes a functional gene product. These second mutations that restore the gene's function are called reversions.

Sometimes a reversion actually restores the gene's original DNA sequence. For example, suppose that a forward mutation changes a G-C pair to an A-T pair. The A-T pair is then replicated faithfully from one generation to the next, perpetuated indefinitely unless changed by a second mutation. If, however, a second mutation changes the mutant A-T pair back to the original G-C pair, then this reversion restores the former function of the gene.

However, reversions, do not necessarily require restoration of the gene's original nucleotide sequence. If a second mutation restores the function of a mutant polypeptide by compensating for the first mutation while still leaving it in place, the second mutation is also called a reversion. For example, a single-nucleotide deletion in the reading frame of a gene shifts the reading frame by one nucleotide for all codons downstream from the mutation, altering most of the amino acids from that point on.

A second mutation that inserts a single nucleotide, upstream or downstream from the site of the deletion, restores the normal reading frame after the ribosome has passed both mutations. Only the codons between the two mutations remain out of frame. If the altered amino acids are not within a critical portion of the polypeptide, the polypeptide's function may be restored, thus qualifying the second mutation as a reversion. Since many nucleotides within a gene are essential for the gene to produce a functional product, mutation of any one of these nucleotides may disrupt the function of the gene's product. Reversions, on the other hand, require at least one of a very few specific changes that restore the function of the gene's product. Since there are many more possibilities for a forward mutation of a functional gene than for reversion of a mutant gene, forward mutations are much more common.

A forward mutation may alter the polypeptide product of a gene, thereby eliminating the function of the polypeptide. However, a mutation within another gene may produce a separate product that compensates for the lost activity of the first. In this case, the second mutation is called a suppressor mutation. Among the best studied suppressor mutations are ones that compensate for nonsense mutations. Suppression of premature UAG termination codons by a mutant tRNA in *E. coli* has been well characterised.

A gene that encodes a tyrosine tRNA with the anti-codon 3' AUG 5' mutates, changing the anti-codon to 3' AUC 5'. This mutant tRNA is charged with tyrosine but pairs with UAG termination codons. When a nonsense mutation with a premature UAG termination codon, and the mutant tyrosine tRNA with a 3' AUC 5' anti-codon, are present in the same cell, tyrosine is encoded by the UAG codon, and translation continues beyond the premature termination codon, restoring the functional polypeptide.

Spontaneous mutation

Mutations are either spontaneous or induced. Spontaneous mutations are naturally occurring mutations due to errors in DNA replication or natural chemical reactions in DNA. They are not caused by external agents. Induced mutations are caused by external agents that chemically alter DNA, usually causing some form of DNA damage.

Tautomeric shifts: Spontaneous tautomeric shifts during DNA replication may cause transitions. Bases in the nucleotides of DNA generally exist in stable forms. However, occasionally a base may shift temporarily to a less stable form that alters its pairing properties. These temporary changes in form are called tautomeric shifts.

For example, the nucleoticles T and G usually exist in the more stable keto-forms, but may occasionally shift to the less stable enol-forms. When either T or G shifts to the enol form during DNA replication, an unusual G-T pair may appear. Likewise, C and A generally assume the more stable amino forms but may occasionally shift to the less stable imino-forms. When either C or A assumes the imino-form during DNA replication, an unusual C-A pair may appear.

Spontaneous tautomeric shifts may cause transitions. For example, a T in the template strand may undergo a tautomeric shift from the keto to the enol form during DNA replication. If DNA polymerase encounters the T in enol form, the polymerase inserts a G opposite the T, instead of an A, thus substituting G (a purine) for A (also a purine).

Mutations in repeated DNA segments: Mutation hotspots are the sites in DNA molecules where insertion and deletion mutations are unusually frequent. DNA sequencing of several mutation hotspots revealed that short, repeated DNA sequences were present at these sites. A repeated DNA sequence may be a mononucleotide repeat (such as AAAAA), a dinucleotide repeat (such as CTCTCTCTCT), a trinucleotide repeat (such as CGGCGGCGGCGGCGG), or longer repeats.

During DNA replication, part of the newly synthesised repeated sequence may slip so that some of the nucleotides in the repeat no longer pair with the template strand. The unpaired part of the newly synthesised strand then loops out from the DNA double helix. Nucleotides of the repeat beyond the loop in the newly synthesised strand may then pair with nucleotides in the template strand at the point of slippage. These nucleoticles are said to be misaligned with the template strand.

DNA polymerase continues synthesis, adding as many extra nucleotides as are present in the loop to the newly synthesised strand. The loop is an insertion mutation. The template strand, rather than the newly synthesised strand, may slip and loop at repeated sequences. In this case, after DNA replication, the newly synthesised strand ends up missing as many nucleotides as are present in the loop of the template strand; in other words, it has a deletion mutation.

Spontaneous lesions: In DNA, chemical reactions may modify a nucleotide in such a way that it no longer carries one of the four standard bases (T, C, A, or G). Such modifications are called lesions. DNA that contains lesions is called damaged DNA. In contrast with lesions, mutations are substitutions, deletions, or insertions of one or more of the four standard nucleotides. A lesion is not a mutation because it cannot be propagated by DNA replication. However, a lesion may cause a mutation when DNA polymerase encounters a lesion in the template DNA strand during DNA replication and incorporates an incorrect nucleotide into the newly synthesised DNA strand. Spontaneous lesions are natural chemical alterations of the bases in the nucleotides of DNA. Depurination is the most frequent cause of spontaneous lesions. In depurination the bond between deoxyribose sugar and a purine base (A or G) is hydrolysed, releasing the base from the DNA backbone and leaving a hydroxyl group in its place. There is then a sugar but no base at this nucleotide position.

Depyrimidination, produced by hydrolysis of the bond between deoxyribose sugar and a pyrimidine base (C or T), is less frequent than depurination but produces the same result: a hydroxyl group on the 1' carbon where a base was previously present. The sites where a base is missing are called apurinic sites or apyrimidinic sites, depending on whether a purine or a pyrimidine was originally present.

As apurinic and apyrimidinic sites are chemically identical, they are called AP sites. With no base present at an AP site in the template strand, there is nothing to which a newly incorporated nucleotide may pair during DNA replication. Consequently, DNA replication ceases at an AP site.

Several repair systems are known that either excise the AP site and replace it with new nucleotides, or allow DNA polymerase to bypass the AP site during replication.

Cytosine deamination is another important spontaneous lesion, in which the amino group on the number 2 carbon of C is removed and a double-bonded oxygen is left in its place, which converts C to U. Since U is not normally found in DNA, it may be recognised by repair mechanisms as an error and be repaired. If left unrepaired, the U pairs with A instead of G during DNA replication, causing a transition.

Cytosines may often be methylated at the number 5 carbon in the base, forming 5-methylcytosine. When 5-methylcytosine is spontaneously deaminated, it becomes a T rather than a U. As just explained, repair systems may recognise U as a lesion in the DNA and correct it. However, because T is one of the four standard nucleotides, repair mechanisms do not recognise it as a lesion, so that, deamination of 5' methylcytosine is a C $\rightarrow$ T transition.

Induced mutations

When cells are exposed to certain external agents, either chemical or in the form of radiation, the rate of mutation may increase. Agents that enter the cell and cause mutations are called mutagens. Mutations that are caused by mutagens are induced mutations. All organisms on earth are exposed to radiation from space and to some naturally occurring chemical mutagens.

Naturally occurring radiation and chemical mutagens have played a significant role in creating new genetic variation by mutation throughout the history of life on earth. A wide variety of agents may cause DNA to mutate. Some agents, such as ultraviolet (UV) radiation in sunlight and ionising radiation from space, are naturally present. Other mutagenic agents are the result of human activity. Mutagenic agents can be divided into two general classes: radiation and chemical mutagens.

Radiation: There are mainly two types of mutagenic radiation: ionising radiation and nonionising radiation.

Ionising radiation is high-energy radiation. Ionising protons, neutrons, X-rays, and alpha, beta, and gamma rays. Ionising radiation may damage DNA directly when high-energy particles strike DNA molecules. However, indirect damage of DNA is a much more common effect of ionising radiation.

When high-energy rays collide with molecules, they often strip atoms of an electron, causing the molecule to be ionised—hence the name ionising radiation. The loss of an electron leaves an unpaired electron within the molecule, which causes the molecule to be highly reactive with other molecules. A highly reactive molecule with an impaired electron is called a free radical. Free radicals may create new free radicals as they collide with other molecules. Free radicals tend to be highly reactive with DNA, attacking and altering the bases. Of even greater consequence, free radicals may attack the phosphodiester bonds in DNA, causing them to break). A break in one strand of double-stranded DNA is generally inconsequential because it may be readily repaired, but when both strands break, repair often fails to restore the DNA molecule to what it was before.

Since their ability to pass through living tissue, X-rays have proved particularly useful in medical and dental diagnoses; the X-rays pass more easily through some tissues than others, providing an image of internal tissues when the X-rays strike radiographic film. For X-ray photographs, radiation is usually brief and at a low level to minimise the risk of radiation-induced disease.

The mutagenic effect of ionising radiation is cumulative, meaning that repeated exposure to low levels of ionising radiation can be dangerous over a period of years. This implies that there is no safe level of radiation, particularly for long-term exposure. A short-term exposure to low-level radiation directed at a limited area, such as an X-ray examination for medical purposes, does not pose a significant danger of harmful mutation, because the exposure is very brief. However, long-term exposure to low levels can be dangerous. For this reason, X-ray technicians use lead aprons and shielding booths to protect themselves from X-ray exposure.

Non-ionising radiation does not have enough energy to strip electrons from atoms. Instead, it causes electrons to assume temporarily higher energy levels within the atom, a phenomenon called excitation.

Ultraviolet (UV) radiation is the most important type of mutagenic nonionising radiation. With its relatively low energy, non-ionising radiation penetrates only the upper most layers of cells in living tissue. UV light is readily absorbed by the nucleotides of DNA, with maximum absorption and maximum mutagenesis. The pyrimidines, C and T, are particularly susceptible to excitation and damage. UV light may convert single pyrimidines into pyrimidine hydrates, lesions that have altered base-pairing properties.

UV light causes covalent bonding between the two adjacent pyrimidines forming a lesion called a pyrimdine dimer. Thymines in particular are highly susceptible to UV-induced dimerisation; the most common dimer, and the most common lesion caused by UV light, is the thymine dimer. If left unrepaired, a pyrimidine dimer in the template strand can cause mutations in the newly synthesised strand during DNA replication.

Chemical mutagens: Chemical mutagens cause lesions in DNA, which, in turn cause mutation when the DNA is replicated. The commonly used chemical mutagens include base analogues, intercalating agents, demainating agents, hydroxylating agents, aflatoxin B_1, alkylating agents and so on.

The nitrogenous bases that are similar to the four bases of DNA are known as base analogs. When present in the form of deoxyribonucleotides, they may be incorporated into DNA during replication in place of the usual nucleotides. Base analogs in a DNA molecule are lesions. The pairing properties of base analogs differ from those of the bases they replace, which mutates nucleotides in the newly synthesised strand during DNA replication when the template strand contains a base analog.

Two common base analogs are 5-bromouracil (5-BU) and 2-amino purine (2-AP).

— *5-BU*: It is an analog of T and, when present during DNA replication, is usually incorporated into DNA in place of T. Like T, 5-BU pairs with A when in the more stable keto form. However, the keto form of 5-BU is much less stable than the keto form of T, resulting in frequent tautomeric shifts of 5-BU from the keto to the enol form. In the enol form, 5-BU (like T in the enol form) pairs with G. Thus, when 5-BU is in the enol form during DNA replication, it causes a transition (A → G) in the newly synthesised strand.

— *2-AP*: It is an analog of A and undergoes frequent tautomeric shifts. Depending on its tautomeric form, it may base-pair with either T or C. Thus, when 2-AP is in the template strand, either T or C may be added opposite it, causing a transition when it pairs with C.

The intercalating agents, are similar to the nitrogenous bases in DNA and RNA. They are stacked flat at right angles to the sugarphosphate backbone. They insert themselves between the stacked bases, a process called intercalation, which causes the DNA molecule to stretch at the insertion point.

If an intercalating agent inserts itself into a single-stranded DNA molecule during replication, it may cause an insertion or deletion mutation, usually of a single nucleotide, causing a frameshift mutation if the mutation is in the reading frame. Intercalating agents include the acridine dyes, such as acridine orange and proflaven, ethidium bromide, and a large class of compounds known as ICR compounds. These agents may intercalate between the bases of both double- and single-stranded DNA and RNA.

Deaminating agents include bisulphite compounds and nitrous acid. Bisulphite ions deaminate C to U, with the same results as spontaneous cytosine deamination. The mutagenic agent nitrous acid (HNO_2) has the ability to deaminate C to U (which pairs with A instead of G), A to hypoxanthine (which pairs with C instead of T), and G to xanthine (which, like G, pairs with C). Consequently, nitrous acid-induced deamination of C and A can cause transitions during DNA replication.

The best studied hydroxylating agent is hydroxylamine, which has a very specific mutagenic effect of adding a hydroxyl group to the amino group attached to the number 4 carbon of C, converting C into hydroxyaminocytosine. The addition of the hydroxyl group causes hydroxyaminocytosine to pair with A instead of G.

Another powerful mutagen, aflatoxin B_1 is attached to guanine and eventually removes the guanine base from the nucleotide, leaving an AP site. In most cases, DNA polymerase inserts an A into the newly synthesised strand across from the AP site, a transversion. Aflatoxin B_1 is a member of a class of large mutagenic molecules that attach to the bases of DNA nucleotides. In some cases, the modified base is hydrolysed from the DNA, creating an AP site.

Alkylating agents add alkyl groups to several positions on all of the nitrogenous bases of DNA, creating lesions. DNA polymeraseses may incorporate incorrect nucleotides in the newly synthesised strand when they encounter alkylated bases in the template strand during DNA, replication. Most mutations caused by alkylating agents are transitions, although these agents are also known to cause transversions, insertions, deletions, and chromosome breaks.

Mutations in certain genes that regulate cell growth and division, are the causing factors of cancer. Consequently, mutagens are often carcinogens (cancer-causing agents).

GENE EXPRESSION

Mutations create variation in the nucleotide sequences of genes and the proteins they encode. Variation in the nucleotide sequences of genes is ultimately expressed as phenotypic variation, the differences in observed characteristics among individuals.

Anatomy of Gene Expression: The Human β-globin Gene

The human β-globin gene provides an excellent example of how a eukaryotic gene is organised and expressed, and how mutations in the gene affect the phenotype. This gene encodes the protein β–globin, which is a subunit of adult hemoglobin. Hemoglobin is a protein essential for the cells of the body to carry out aerobic respiration. It is located in erythrocytes and is responsible for binding oxygen and carrying it from the lungs to the various tissues of the body. In the tissues, erythrocytes enter capillaries, tiny blood vessels that allow erythrocytes to pass through one at a time. In the capillaries, hemoglobin delivers oxygen to the tissues and binds carbon dioxide. Hemoglobin then carries the carbon dioxide to the lungs, where it exchanges the carbon dioxide for oxygen.

Since hemoglobin provides an essential function for all tissues in the body, mutations that alter hemoglobin function can cause widespread organ and tissue damage, serious illness, and death. The locations of conserved sequences and the codons for the polypeptide can identified in the DNA sequence. The entire gene region, including the promoter, is about 1700 nucleotide pairs in length and includes two introns that constitute about 60 per cent of the nucleotides in the gene. This gene is among the smallest protein-encoding genes in humans, making it a simple one to study.

Promoter of the β-globin gene is a good example 41 of a eukaryotic promoter. It contains two CACCC boxes, one at—106 and a second at —90. The CCAAT box is located at —75, and the TATA box at —29. The CACCC and CCAAT boxes match their consensus sequences exactly, whereas the TATA box deviates from its consensus sequence by one nucleotide (CATAAAA, compared to the consensus sequence TATAAAA).

The transcription startpoint is an A between two pyrimidines. After it has been transcribed, the RNA is cleaved 18 nucleotides beyond a fully conserved AAUAAA site, yielding a pre-mRNA that is 1604 nucleotides in length. The A at the 5' end of the pre-mRNA is capped, and a poly (A) tail is added to the 3' end. There are two introns in the pre-mRNA. The first intron is 130 nucleotides in length, and the second is 850 nucleotides in length. Both introns are within the reading frame, so precise intron removal is essential to prevent a frameshift. The introns adhere to the GT-AG rule. The 5' splice site of the first intron contains the conserved sequence GTTGGT, which differs by two nucleotides from the GTAAGT consensus sequence. The 3' splice site contains the conserved sequence CCACCCTTAG, which differs by two nucleotides from the 6PyNCAG consensus sequence.

The sequence CACTGAC, which corresponds to the consensus sequence PyNPyPyPuAPy, is 39 nucleotides upstream from the 3' splice site. The second intron has similar conserved sequences. At the 5' splice site is the sequence GTGAGT, which differs by one nucleotide from the consensus sequence. At the 3' splice site is the sequence CCTCCCACAG, which matches its 6PyNCAG consensus sequence perfectly.

The sequence TGCTAAT lies 35 nucleotides upstream from the 3' splice site and corresponds to the PyNPyPyPuAPy consensus sequence. Once the introns are removed, the mature mRNA is 624 nucleotides long, not counting the poly (A) tail. The reading frame begins with the first AUG in the mRNA, which is located 50 nucleotides downstream from the transcription startpoint.

The AUG initiation codon lies within the sequence ACAGACACCAUGG, which differs from the GCCGCCPuCCAUGG translation initiation consensus sequence by three nucleotides. The purine at –3 from the AUG codon and the G at +1 from the AUG codon, which (except for the AUG) are the two most important nucleotides in the consensus sequence, are fully conserved. The first in-frame stop codon is UAA, which is the 148th codon in the reading frame, yielding a polypeptide that is 147 amino acids in length. Following translation, the methionine at the amino end of the polypeptide is removed, leaving 146 amino acids in the primary structure.

Most of the amino acids coil into eight helices that fold around a single heme group to form the tertiary structure. Each heme group consists of an iron atom surrounded by a protoporphyrin ring. The heme groups are examples of a nonprotein substance added to each protein subunit to form the tertiary structure during post-translational modification of the protein. Once it has assumed its tertiary structure, α, β subunit binds to an α subunit to make a dimer. Two dimers then attach to each other to form the tetrameric hemoglobin protein.

Hemoglobin carries oxvgen from the lungs through the bloodstream to the tissues and releases oxygen when the erythrocyte arrives in a capillary. As it releases oxygen, the hemoglobin molecule picks up carbon dioxide and carries it through the bloodstream to the lungs, where the carbon dioxide is expelled. Thus, hemoglobin must be capable of alternatively binding oxygen or carbon dioxide. Hemoglobin binds oxygen by drawing an O_2 molecule into a pocket located in each of the four subunits. The pocket is called the heme pocket because a heme group resides within it.

Heme groups are hydrophobic, which is essential for oxygen binding. The pocket is lined with hydrophobic amino acids that prevent water from entering the heme pocket. The path into the pocket is just large enough for an O_2 molecule to penetrate and bind to the heme group. The amino acids on the outer surface of the hemoglobin molecule are mostly hydrophilic, making the hemoglobin molecule soluble in water. This is an effective way to allow a water-insoluble heme group to function in the aqueous interior of an erythrocyte. When hemoglobin encounters carbon dioxide in the capillaries, arginines located near the carboxyl end of the α-globin subunits bind carbon dioxide.

Even though the binding sites for oxygen and carbon dioxide are different, hemoglobin does not bind both molecules simultaneously. Carbon dioxide binding alters the conformation of hemoglobin, causing the hemoglobin molecule to lose its affinity for oxygen. This facilitates the exchange of oxygen and carbon dioxide as hemoglobin delivers oxygen to the tissues. When hemoglobin is in the lungs, it releases carbon dioxide, which increases hemoglobin's affinity for oxygen.

Disorders Caused by Mutation of β-globin Gene

A large number of mutations have been identified in the human β-globin gene. Some of these mutations are especially frequent among people of certain ancestral groups, making hemoglobin disorders among the most important genetic disorders throughout the world.

Sickle-cell anaemia

Sickle-cell anaemia is characterised by weakness, fatigue, heart failure, joint and muscular impairment, abdominal pain and dysfunction, impaired mental function, and eventual death, often during childhood. The mutation that causes sickle-cell anaemia in humans was among the first mutations characterised at the molecular level in any organism.

The sixth amino acid in the mature, β-globin polypeptide is glutamic acid, which is encoded by GAG. An A → T substitution in the second nucleotide of the codon changes the codon from GAG to GUG, substituting valine for glutamic acid in the polypeptide. A normal hemoglobin molecule, with glutamic acid at position 6 in the β subunits, is called Hemoglobin A (Hb A), whereas the mutant molecule, with valine at position 6 in the β subunits, is called Hemoglobin S (HbS). Glutamic acid is acidic and hydrophilic, whereas valine is nonpolar and hydrophobic. This substitution therefore alters the chemical properties of hemoglobin.

The sixth amino acid is not located within the heme pocket, so it does not directly prevent oxygen binding. However, when all the hemoglobin molecules are HbS, the HbS molecules attach to one another when they are deoxygenated, forming long chains. Remember that oxygen is bound in the heme pocket when hemoglobin is oxygenated. When oxygen is not bound, the hydrophobic valine at position 6 embeds itself into the hydrophobic heme pocket in a β subunit of another deoxygenated hemoglobin molecule.

Interactions between several amino acids within the mutant HbS molecules cause them to attach to one another, forming a long double-stranded polymer. Seven of these double-stranded polymers then wind around each other to form a sickle hemoglobin fibre. The HbS molecules in sickle hemoglobin fibres have reduced oxygen affinity and cause erythrocytes to assume a sickle shape instead of the normal disk shape, which is the reason for the name sickle-cell anaemia.

Sickled cells do not pass through capillaries as readily as normal disk-shaped cells, so sickled cells often clog capillaries, denying oxygen to the cells served by them. Lack of oxygen and accumulation of carbon dioxide cause widespread tissue damage throughout the body. As the spleen removes sickled cells, it becomes enlarged and suffers damage.

β–Thalassemia

β-thalassemia is caused by a reduction of β-globin synthesis (called β^+-thalassemia) or the absence of β-globin (called β^0-thalassemia). When a reduction of β-globin synthesis disrupts the 1:1 ratio of a and a subunits, hemoglobin synthesis is impaired and the excess a subunits aggregate and precipitate. This damages erythrocytes and leads to their premature destruction. The phenotypic symptoms of severe b-thalassemia include expansion of bone marrow, which causes skeletal deformities and bone fractures.

The reduction of functional hemoglobin also causes iron overloading, which may lead to cirrhosis of the liver, endocrine dysfunction, diabetes, and cardiac failure. Like patients with sickle-cell anemia, many patients with severe β-thalassemia suffer an early death. Since the amount of functional, β-globin may vary among patients, the severity of symptoms may vary as well. Over 30 different mutations in the b-globin gene are known to cause β-thalassemia.

Studies in human β-globin gene showed how mutations in a gene may alter the protein product, which may in turn alter the phenotype. However, it is erroneous to conclude that a particular mutant phenotype is always caused by mutations in the same gene.

Mutations in any gene that reduces or eliminates hemoglobin function may cause similar symptoms. For example, several mutations in the α–globin gene cause β-thalassemias, which are similar in their phenotypic effects to the β-thalassemias. Also, mutations in the genes that encode enzymes for heme synthesis may reduce or eliminate functional hemoglobin due to alterations in heme production.

Mutations in any of the several genes that participate in hemoglobin synthesis may be expressed as similar phenotypes because all of the mutations have a similar result: the reduction or absence of functional hemoglobin. Most phenotypic characteristics are influenced by the combined effects of several genes, although a mutation in any one gene may alter the phenotype.

Mutations and Biochemical Pathways

In a cell, compounds arise through a series of steps known as a biochemical pathway. Each step is determined and regulated by a specific enzyme, and each enzyme is encoded by one or more genes. The pathway begins with a pathway substrate, a compound that the first enzyme in the pathway converts into another compound, which is called an intermediate. Each intermediate is the product of one enzyme, and is also the substrate for the next enzyme in the pathway. Enzymes continue step by step, converting one intermediate into another until the last intermediate is converted into the end product.

The genes that encode the enzymes of a biochemical pathway function to produce the same end product of the pathway, so that mutations in any one of the genes may cause the same or a similar mutant phenotype. The idea that genes were expressed as enzymes dates back to just after 1900. Evidence available in the latter part of the nineteenth century indicated that biochemical pathways were controlled by enzymes.

In 1902, Archibald Garrod recognised the relationship between the absence of a functional enzyme and an inherited disorder, implying that genes influenced enzymes. Garrod had been studying a disorder in humans called alkaptonuria, which he had identified as an inherited disorder on the basis of pedigree analysis. This disorder can be easily diagnosed in affected individuals because their urine turns dark when exposed to the air.

Garrod identified a substance, called alkapton in his day, that was present in large amounts in the urine of affected individuals but in only trace amounts in that of unaffected individuals. We now call this substance homogentisic acid. When exposed to oxygen, homogentisic acid turns a dark colour, causing the evacuated urine of people with alkaptonuria to darken. Garrod concluded that the excess of homogentisic acid in the urine of people with alkaptonuria was due to the absence of an enzyme that normally degrades homogentisic acid. When this enzyme is absent, homogentisic acid accumulates because there is nothing to degrade it.

A mutation in the gene that encodes the enzyme homogentisic acid oxidase eliminates production of the functional enzyme, blocking the pathway and causing accumulation of homogentisic acid. We can compare a biochemical pathway to a street and the enzymes controlling the flow of the pathway to the stoplights on that street. The absence of a functional enzyme is like a defective stoplight that fails to turn green: traffic passes through all the stoplights that are functioning normally, but backs up behind the defective one.

Mutations in different genes sometimes confer the same phenotype. In many cases, these different genes encode different enzymes that participate in the same biochemical pathway. For example, mutations in several different genes in *Drosophila melanogaster* confer a brilliant-red colour to the eyes instead of the usual brick-red colour. A classic study by George Beadle and Boris Ephrussi, summarised in the following example, demonstrated that two of these genes encode different enzymes in the same pathway.

Beadle, together with Edward Tatum, did experiments with fungi of the genus *Neurospora*, a group of bread-mould fungi that can be easily grown in culture plates and studied genetically. *Neurospora* can be grown on a simple medium called minimal medium, composed of a sugar, several minerals, and the vitamin biotin. The cells are able to synthesise all other molecules they require from the substances in the minimal medium.

Beadle and Tatum treated cultures of *Neurospora* with X-rays to induce mutations. They then recovered three different mutant strains that could not grow on minimal medium but could grow when the medium was supplemented with malt extract and yeast extract. They concluded that substances not present in the minimal medium were present in the extracts and that these substances allowed the mutant strains to grow. This type of mutation, which confers a requirement for supplementation of minimal medium to overcome the effects of the mutation, is called an auxotrophic mutation. By supplementing minimal medium with individual vitamins and amino acids, Beadle and Tatum determined which substances overcame the effects of each auxotrophic mutation. At the time, it was known that biochemical reactions were mediated by enzymes. However, many scientists believed that genes acted directly on the phenotype, rather than through enzymes.

From their experimental results, Beadle and Tatum concluded that a mutation in one gene eliminated the function of one enzyme, which led them to propose the one gene-one enzyme hypothesis, the idea that each gene produced one enzyme. The one gene-one enzyme hypothesis constituted a major shift in thinking. Beadle and Tatum proposed that, instead of a gene acting directly on the phenotype, genes produce enzymes that catalyse biochemical reactions, and the products of these reactions determine the phenotype. Researchers later discovered that some enzymes consist of two or more protein subunits that together make the functional enzyme, implying that some enzymes are encoded by two or more genes. Also, there are many proteins that are not enzymes. For these reasons, the one-gene-one-enzyme hypothesis was modified to the one-gene-one-polypeptide hypothesis, which reflects more accurately the relationship of genes and their products. Beadle and Tatum received a Nobel

Prize in 1958 for their work in discovering the relationship of genes and enzymes.

REFERENCES

Alberts, B., D. Bray, J. Lewis, M. Raff, K. Roberts, and J. D. Watson., 1994, *Molecular Biology of the Cell*, 3rd ed. New York: Garland.

Beadle, G. W., and B. Ephrussi, 1937, Development of eye colours in Drosophila Diffusible substances and their interrelations. *Genetics* 22:76-86.

Beadle, G. W., and E. L. Tatum, 1941, Genetic control of biochemical reactions in *Neurospora: Proceedings of the National Academy of Sciences*, USA 27:499-506.

Collins, F., and S. M. Weissman, 1984, The molecular genetics of human haemoglobin. *Progress in Nucleic Acid Research and Molecular Biology* 31:315-462.

Cook-Deegan, R. M., 1994, *The Gene Wars: Science, Politics, and the Human Genome*, New York: Norton.

Crick, E. H. C., 1962, 'The genetic code', *Scientific American* 207 (Oct): 66-74; Nirenberg.

Eisensmith, R. C., and S. L. C. Woo, 1995, 'Molecular genetics of phenylketonuria: From molecular anthropology to gene therapy', *Advances in Genetics* 32:199-271.

Handin, R. L, S. E. Lux, and T. P. Stossel, (eds.), 1995, *Blood: Principles and Practice of Haematology*. Philadelphia: Lippincott.

Merrick, W. C., 1992, Mechanism and regulation of eukaryotic protein synthesis. *Microbiological Reviews* 56:291-315.

Moore, J. A., 1993, *Science as a Way of Knowing: The Foundations of Modern Biology*, Cambridge, Mass: Harvard University Press.

Olby, R. C., 1985, *Origins of Mendelism*, 2nd ed. Chicago: University of Chicago Press.

Parker, J. 1989, Errors and alternatives in reading the universal genetic code. *Microbiological Reviews* 53:273-298.

Prota, G., 1992, *Melanins and Melanogenesis*. San Diego: Academic Press.

Sturtevant, A. H., 1965, *A History of Genetics*. New York: Harper & Row.

Woods, R. A., 1980, *Biochemical Genetics*, 2nd ed. London: Chapman and Hall.

3

DNA: The Genetic Material

The deoxyribonucleic acid (DNA) is the genetic material of all cellular organisms ranks as one of the greatest accomplishments in science. We usually think of the discovery of DNA as something recent, and indeed DNAs structure, mode of replication, and function were not known until the 1950s. However, the actual discovery of DNA as a substance dates back to the 1860s, about the same time as Gregor Mendel's discoveries. With refinements in microscopy, cells and their components were under intensive study in the 1860s. The most obvious cellular component was the nucleus, and for this reason a number of scientists focused their research efforts on it. In 1866, Ernst Haeckel proposed that the nucleus contains the factors necessary for heredity, setting the stage for determining what substances are present in the nucleus.

In 1860s, Friedrich Miescher began physiological and chemical studies of pus cells (simple human cells that were easily available because antiseptic medicine was not often practised and postsurgical infections were common). Miescher intended to characterise substances called proteins. Proteins had been discovered about 30 years earlier and were thought to be the most important material in the cell. When Miescher separated the nuclei from the rest of the cell components, he discovered that the nuclei had large quantities of a substance that did not behave chemically like protein. This substance was eventually named deoxyribonucleic acid, which is usually abbreviated to DNA. According to Miescher DNA was a reservoir of cellular phosphorus made available on demand by digestion. Miescher studied DNA from salmon sperm and reported an empirical formula for it ($C_{29}H_{49}N_9P_3O_{22}$) that is fairly close to the actual proportional composition of these elements in DNA.

In the nineteenth century, Edmund Wilson used evidence from the staining properties of DNA to conclude that there is considerable ground for the hypothesis that in a chemical sense this substance is the most essential nuclear element handed on from cell to cell, whether by cell division or fertilisation. Thus, before the rediscovery of Mendel's principles of inheritance in 1900, DNA had already been postu-lated as the genetic material.

However, Wilson later reversed his view, stating that it was nuclear protein that persists through cell division and that DNA appeared and disappeared during the various stages of a cell's life. With the discovery and characterisation of chromosomes, it became increasingly clear that chromosomes are inherited and that they are composed of chromatin, a combination of both protein and DNA. The view that the protein in chromatin, rather than the DNA, is the hereditary material persisted until the middle of the twentieth century.

TRANSFORMING PRINCIPLE

During 1928, Frederick Griffith identified the genetic material in bacteria. He was a US Army medical officer attempting to develop a vaccine against pneumonia. He studied the infection patterns of the pneumonia-causing bacterium *Streptococcus pneumoniae* using mice as the experimental host organisms.

Griffith studied two types of the bacterium, one with a polysaccharide on its capsular surface that prevents it from being destroyed by the host's immune system, and one that is unencapsulated and lacks the polysaccharide. The host's immune system easily destroys the unencapsulated cells. The bacteria with the polysaccharide have a smooth appearance when grown as colonies on a petriplate and are called S for 'smooth', whereas the bacteria-lacking the polysaccharide form colonies with a rough appearance and are called R for 'rough'.

The S-type bacteria are virulent (disease causing), and the R-type bacteria are avirulent (harmless). When Griffith injected live S-type bacteria into a mouse, the mouse died from pneumonia. But if he heat-killed the S-type bacteria before injecting them, then the mouse did not die, presumably because the dead S-type bacteria could no longer reproduce.

When Griffith injected live R-type bacteria into a mouse, the mouse did not die because the mouse's immune system destroyed the R-type bacteria. However, when Griffith injected heat-killed S-type bacteria together with live R-type bacteria into a mouse, the mouse died and he found live S-type bacteria in the mouse's blood. From his experiments Griffith concluded that something in the dead S-type bacteria is not destroyed by the heat treatment and is transmitted to the live R-type bacteria, transforming them into S-type bacteria. Moreover, whatever is transmitted must be inherited because he observed that the descendants of the cells transformed from R into S were also S. Griffith called the transmitted substance the transforming principle.

During 1944, Oswald Avery, Colin MacLeod, and Maclyn McCarty reported a follow-up to Griffith's experiments in which they sought to identify Griffith's transforming principle. They heat-killed S cells, then treated the cells chemically to purify a viscous substance that transformed R cells into S cells. They specifically treated the substance to remove protein and RNA. Their chemical analysis of this substance revealed that its composition matched DNA. There remained, however, the possibility that the substance was contaminated with small amounts of protein or RNA, and that the contaminant, rather than the DNA, was responsible for the transformation. When they treated the substance with deoxyribonucleases (DNA-degrading enzymes), they discovered that it no longer transformed R cells into S cells.

However, if they treated the substance with proteases (protein-degrading enzymes) or ribonucleases (RNA-degrading enzymes), the substance retained its transforming ability. These experiments ruled out the possibility that a protein or RNA contaminant of the DNA preparation might actually be the transforming principle.

Avery, MacLeod, and McCarty determined that within the limits of the methods, the active fraction contains no demonstrable protein, unbound lipid, or serologically reactive polysaccharide and consists principally, if not solely, of a highly polymerised, viscous form of deoxyribonucleic acid (DNA). Their experiments demonstrated quite conclusively that Griffith's transforming principle is DNA.

NUCLEIC ACID STRUCTURE

DNA is classified as a nucleic acid, as is ribonucleic acid, or RNA. They are called nucleic because much of the material found in the cell nucleus is DNA and RNA. And they are referred to as acids because they tend to be acidic when dissolved in aqueous solution. DNA is the genetic material of all cellular organisms. RNA plays crucial roles in gene expression and in DNA replication in all organisms, and many viruses have RNA instead of DNA as their genetic material.

DNA Molecule

Compared to most other molecules, DNA is enormous in size. Although we measure the size of most molecules in angstroms (one angstrom, A, equals one ten-billionth of a metre, 10^{-10} m), we measure certain DNA molecules in millimetres. Human beings have 46 separate DNA molecules in the nucleus of each cells. These DNA molecules, if stretched out end-to-end, are on average about 50 mm long, about the length of your little finger. If all the DNA molecules in the nucleus of one of your cells were lined up end-to-end, the total length would be over 2 meters.

DNA contains enormous amount of information in it. The DNA in the nucleus of one diploid human cell consists of about six billion pairs of subunits. If a single letter the size of the letters in this sentence represented each subunit pair, and if 6 billion of those letters were written out in a single line, it would extend for over 6, 600 km.

A DNA molecule is composed of four subunits called nucleotides. Each nucleotide consists of a five-carbon sugar, deoxyribose, with a nitrogenous base attached to one end and a phosphate group attached to the other. In the five-carbon deoxyribose sugar, the first four of the five carbon atoms, together with an oxygen atom, form a five-cornered ring, with the fifth carbon extending out from the ring. The five carbon atoms are designated 1', 2', 3', 4', and 5' (pronounced "one-prime," "two-prime," "three-prime," etc.). The primes distinguish the carbons of the sugar from the numbered carbon and nitrogens in the nitrogenous base.

The nitrogenous base is attached to the 3' carbon of the sugar, and a phosphate group is attached to the 5' carbon. An oxygen atom that is normally present on the 2' carbon in the ribose sugar is absent in DNA, hence the prefix deoxy in the term deoxyribonucleic acid.

There are four different nucleotides in DNA identical in structure except for the composition of the nitrogenous bases. The four nitrogenous bases are thymine, cytosine, adenine, and guanine. The chemical names of the nucleotides are: deoxythymidine 5'-monophosphate (dTMP), deoxycytidine 5'-monophosphate (dCMP), deoxyadenosine 5'monophosphate (dAMP), and deoxyguanosine 5'monophosphate (dGMP) respectively. Each nucleotide in a DNA molecule is usually referred to by the name of its base, and often by the first letter of its base: T, C, A, or G. C and T, which are called pyrimidines, are similar to each other in structure, each having a single six-membered ring in the nitrogenous base. The other two nucleotides, A and G, called purines, are also similar to each other in structure; each purine has a nitrogenous base composed of two rings, a six-membered ring fused with a five-membered ring.

RNA Molecule

RNA is composed of ribonucleotides that differ only slightly from their deoxyribonucleoticle counterparts in DNA, Each ribonucleotide in RNA consists of a five-carbon ribose sugar that, as in the deoxyribonucleotides, has a nitrogenous base attached to its 1' carbon and a phosphate group attached to its 5' carbon. Ribonucleotides, however, have a hydroxyl group attached to the 2' carbon, whereas deoxyribonucleotides have only a hydrogen in this position.

Like DNA, RNA consists of four nucleotides that differ from each other only in the composition of the nitrogenous base. Three of the nitrogenous bases are identical to three of the bases in DNA-cytosine, adenine, and guanine. The fourth nitrogenous base in RNA, uracil (U), is very similar to thymine in DNA. Thymine has a methyl group attached to carbon number 5, but in uracil this methyl group is missing.

DNA and RNA are polymers. The polymers form as the phosphate group at the 5' end of one nucleotide attaches to the 3' carbon of another nucleotide. We call the attachments phosphodiester bonds because there are two ester linkages, one on either side of the phosphorus atom. These phosphodiester bonds connect one nucleotide to the next to form a chain of nucleotides called a single strand of DNA or RNA. The 5' carbons of each nucleotide in a single strand all point in the same direction. At one end a strand has a 5' carbon with a phosphate group attached, and the other end terminates with a 3' carbon attached to a hydroxyl group.

DNA and RNA also exist in double-stranded forms in which two single strands of nucleotides wind around each other to form a double helix. Most DNA molecules are double stranded, whereas most RNA molecules are single stranded. The nucleotides T, C, A, G, and U connect to each other with phosphodiester bonds to form the sugar-phosphate backbones of nucleic acids.

Watson and Crick Model of DNA

During 1951, James Watson and Francis Crick, recognised that experiments such as those by Hershey and Chase, and Avery, MacLeod, and McCarty provided good evidence that DNA was the hereditary material, and that discovery of its structure would therefore be of great biological importance. Watson and Crick were fortunate to have access to X-ray diffraction patterns of DNA crystals provided by Rosalind Franklin and Maurice Wilkins.

Combining the X-ray observations with additional information, Watson and Crick developed a number of models of the structure of DNA in an attempt to find one that matched the experimental evidence and fit the rules imposed by stereochemistry. Early models included a triple-stranded helix, a model in which the phosphate groups pointed toward the centre of the helix, and a model in which individual bases paired with themselves, such as adenine with adenine, guanine with guanine, and so on.

The X-ray images taken by Franklin predicted that the DNA molecule is a helix of uniform width that repeats its structure every 34 angstroms and has the bases pointing inward. Chargaff's data predicted that adenine and thymine are present in equal amounts, and that guanine and cytosine are present in equal amounts. Each of the early models failed to meet at least one of these restrictions and had problems with rules of chemistry.

However, the final model, in which adenine paired with thymine and guanine paired with cytosine, agreed with the experimental data and the rules of stereochemistry. Once Watson and Crick revealed their model, scientists who had criticised the previous models recognised that this one was correct. Watson and Crick published their results in 1953 in what is now a classic paper in genetics.

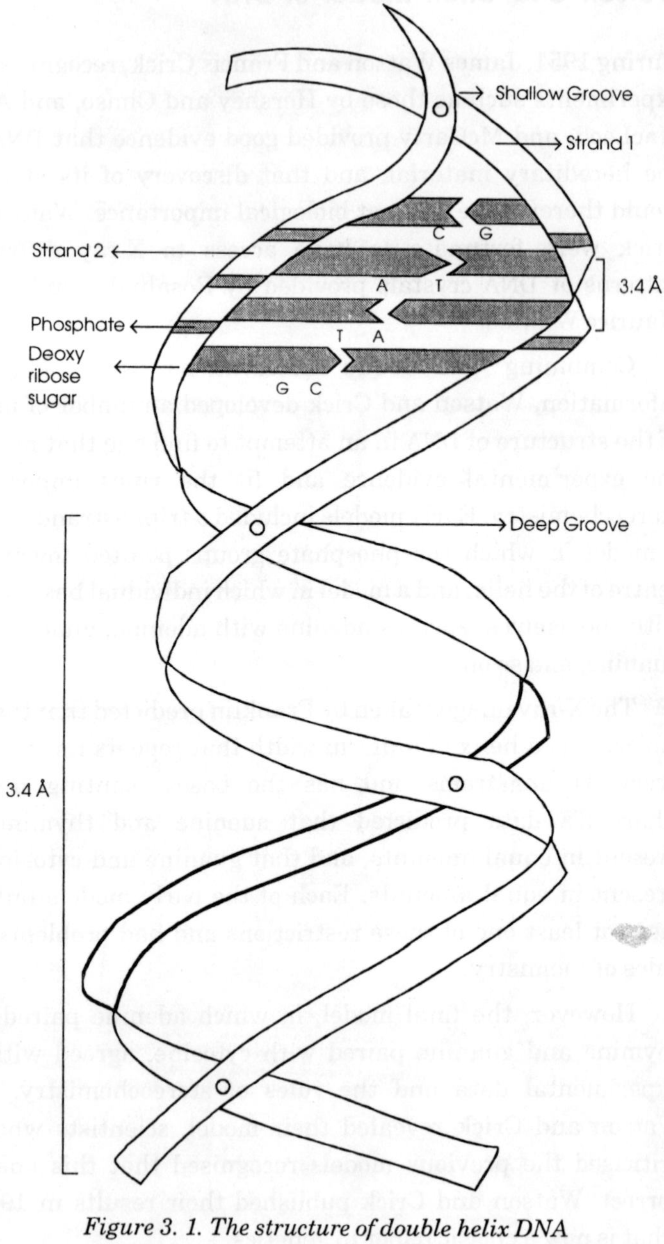

Figure 3. 1. The structure of double helix DNA

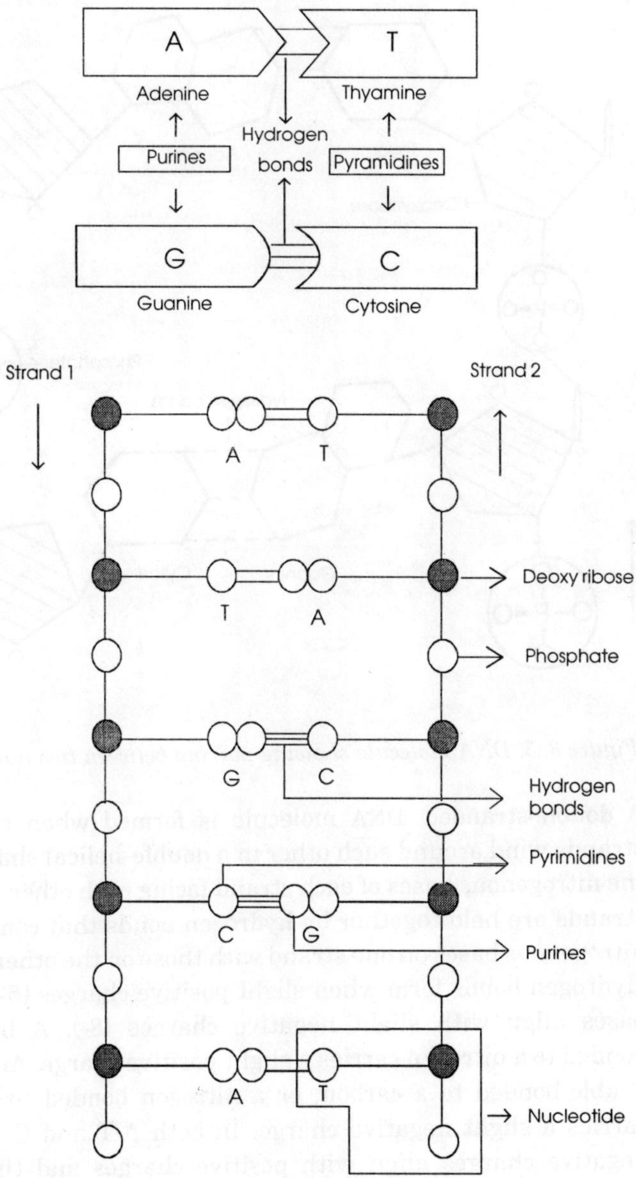

Figure 3. 2. The structure of DNA molecules

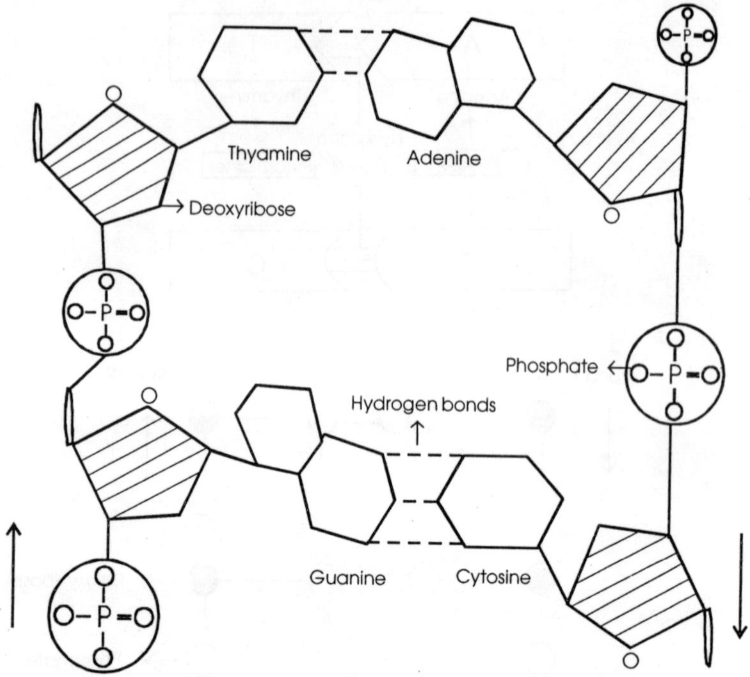

Figure 3. 3. DNA molecule showing pairing between two nucleotides.

A double-stranded DNA molecule is formed when two such strands wind around each other in a double-helical shape, with the nitrogenous bases of each strand facing each other. The two strands are held together by hydrogen bonds that connect the nitrogenous bases on one strand with those on the other strand. Hydrogen bonds form when slight positive charges (5+) in the bases align with slight negative charges (8-). A hydrogen bonded to a nitrogen carries a slight positive charge. An oxygen double-bonded to a carbon, or a nitrogen bonded to carbons carries a slight negative charge. In both A-T and G-C pairs, negative charges align with positive charges and the bases attract each other, forming hydrogen bonds. This results in some very precise nucleotide pairing rules for DNA: T pairs with

A, and C pairs with G. The paired nucleotide sequence of the double-stranded DNA segment would be as follows:

5' AATAGCCA 3'
3' TTATCGGT 5'

The two strands must be anti-parallel, which means that the 5'-3' orientations of the two strands in the double helix run in opposite directions. A close look at nucleotide structure reveals the chemical basis for the base-pairing rules. The nitrogenous bases fit much like the interlocking pieces of a jigsaw puzzle. A pyrimidine must pair with a purine. Remember that purines have two rings in the nitrogenous base, whereas pyrimidines have only one ring, making purines larger than pyrimidines. If a purine paired with another purine, there would be four rings across the axis of the DNA molecule. However, the DNA molecule has a precisely uniform width throughout its length because purines only pair with pyrimidines.

In the Watson-Crick model of double-stranded DNA, the nitrogenous base pairs lie flat against each other, forming planes that are almost perpendicular to the axis of the molecule, much like the stairs of a spiral staircase. The deviation from perpendicular is about 6'. The bases stack tightly against each other. In the stack, the axis of each base pair is rotated approximately 36' relative to the previous base, so the molecule makes one complete turn of 360' about every ten base pairs. The turns of a helix may be either right-handed or left-handed; this orientation is consistent regardless of the direction from which the helix is viewed. For instance, the threads on a typical bolt or wood screw are right-handed, meaning that bolts and screws are tightened by turning them in a clockwise direction. The groove in the threads of a screw is a single helix. You can visualise the single helix by placing the point of a felt-tip pen in the groove at the tip of a screw, then rotating the screw clockwise, holding the pen in the groove until it reaches the head of the screw. The pen will deposit ink in the single groove throughout the entire length of the screw.

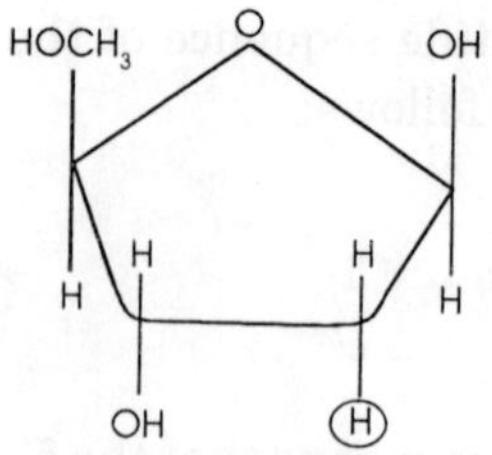

1. Deoxyribose

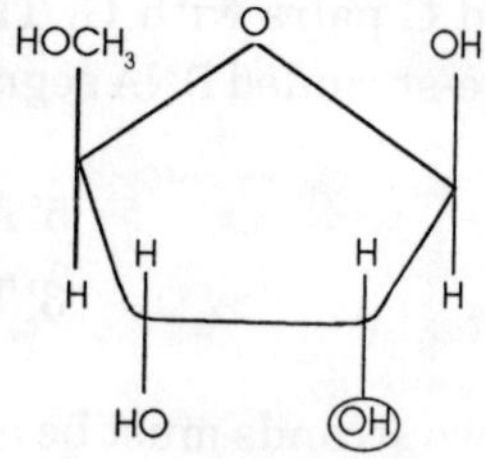

2. Ribose

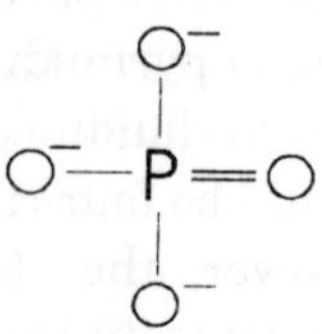

3. Phosphate

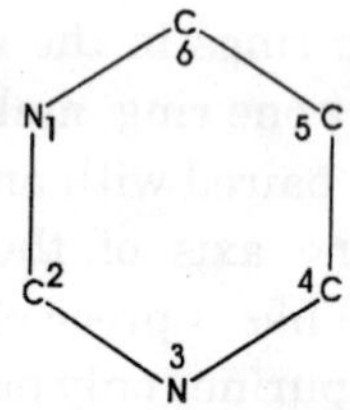

4. Pyrimidene ring

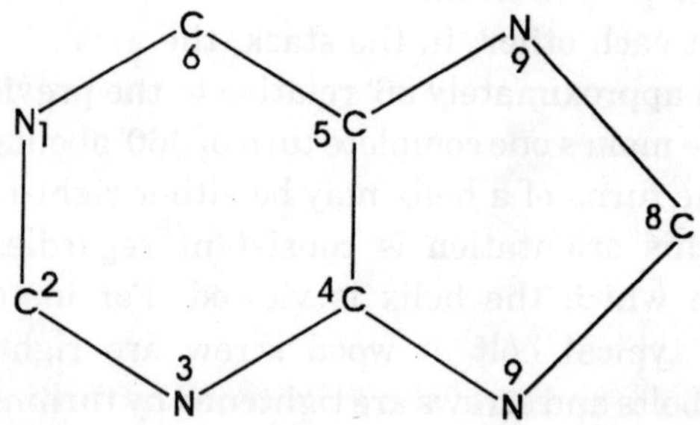

5. Purine ring

Figure 3. 4. Chemical formulas

In Watson-Crick model, DNA is a right-handed double helix. Since it is a double helix, there are two separate grooves that twist around the molecule between the sugar-phosphate backbone, much like the red and white stripes on a barber-shop pole. Looking at the molecule from its side, the two different grooves appear once in each full turn of the helix. The two grooves in DNA are not equal in size. The major groove is approximately 22 Å wide, and the minor groove 12 Å wide, making the linear distance for one complete turn of the helix about 34 Å, with 3.4 Å between each base pair. This form of DNA, with the bases stacked nearly perpendicular to the axis of the molecule at about a 36' angle of rotation for each base pair and with major and minor grooves in a right-handed double helix, is referred to as the B form and is the most common model for DNA.

Forms of DNA

Scientists have detected alternative forms in purified DNA that has been exposed to different conditions of relative humidity and salt concentration. Most alternative forms are found only under conditions not normally present in cells. However, under certain conditions, an alternative form of DNA may appear in certain regions of DNA within living cells. The alternative forms of DNA differ from each other primarily in the degree and direction of helical winding and in the stacking of the bases. The DNA can exist in the A, B, C and D, H, and Z forms.

In A-DNA, the bases are tilted at about 200 from perpendicular to the axis of the molecule. The helix is wound less tightly than in B-DNA, with a 32.7' rotation per base pair. Scientists can induce DNA to assume the A form under laboratory conditions, but A-DNA does not exist in living cells. However, double-stranded DNA-RNA hybrids and double-stranded RNA, which are found in living cells, assume a conformation like A-DNA.

Z-DNA is unusual in that its double helix is wound in a left-handed conformation, the molecule is quite narrow, and there are more base pairs per helical turn than in B-DNA. As viewed from outside the space-filling representation, the sugar-phosphate backbones of Z-DNA form a zig-zag pattern, which is the reason for the Z-designation. Although typical cellular conditions do not favour the formation of Z-DNA, certain situations may cause it to appear within limited regions of the DNA molecule. These may include cases where specialised proteins cause left-handed twisting of the B-DNA molecule or when the nucleotide sequence favours the Z-DNA form. Antibodies specific for Z-DNA bind to a few regions of DNA in cells, suggesting that limited regions of Z-DNA are present.

H-DNA is a triple-helical form of DNA. A short region of double-helical B-DNA may rearrange to become the H-DNA triple helix. There is good evidence that short segments of H-DNA are present in living cells and that the H-DNA may help control the expression of genes.

B-DNA appears to be by far the most common form in living cells. Other forms known as C, D, and E-DNA can be induced to form under laboratory conditions but have not been found in living cells. In a typical DNA molecule, two sugar-phosphate backbones wind together in a double helix with their attached nitrogenous bases pointing toward the centre of the helix. There are several alternative forms of DNA, although B-DNA is the most common in living cells.

Paliandromic DNA

A palindromic DNA is a sentence that read the same forwards and backwards e.g.., 'Madam I am Adam'. The DNA of many eukaryotes are shown to have paliandromic sequence, in which nucleotide of one strand going in one direction are the same as the nucleotides of the other strand going in the same direction, e.g.,

```
—A—A—T—T—G—C—A—A—T—T—
 |  |  |  |  |  |  |  |  |  |
—T—T—A—A—C—G—T—T—A—A—
```

DNA REPLICATION

When Watson and Crick first described the structure of DNA, they recognised that the base pairing suggested the mode of replication. The basic model of DNA replication is quite simple. The two strands 'unzip' by breaking the hydrogen bonds that hold the paired nitrogenous bases together, creating two single-stranded portions of the molecule and exposing the unpaired bases on these portions. The place at which the two strands separate is called the replication fork. Free nucleotides that are present in the cell may then pair with the nucleotides on each of the single strands according to the base-pairing rules, making two identical molecules.

As the molecule unwinds, the replication fork moves farther along until the molecule is fully replicated into two identical double-stranded DNA molecules. This mode of replication is referred to as semi-conservative, meaning that the two original strands are conserved, each serving as a template on which a new strand forms. The two daughter molecules each contain an original parental strand and a newly synthesised strand.

Although the structure of DNA led Watson and Crick to devise the model of semi-conservative replication, there was no experimental evidence to confirm this hypothesis. The Watson-Crick model allowed for other models, and two were proposed. In the conservative replication model, the original double-stranded molecule was conserved, and both strands of the replicated molecule were new. In the dispersive replication model, each strand on both daughter molecules combined parental and newly synthesised DNA.

During 1958, Matthew Meselson and Franklin Stahl published the results of experiments designed to reveal which model is correct. They provided actively dividing *E. coli* cells with a heavy isotope as their source of nitrogen. As the cells grew, they incorporated the 5N into the nitrogenous bases of their DNA. After 17 cell generations, all DNA in the bacteria was labelled with 15N. The labelled bacteria were then transferred to a medium containing only 14 N. Some were allowed to divide once (thus replicating their DNA once), and others were allowed to divide twice (thus replicating their DNA twice). The DNA from these groups of bacteria was isolated and centrifuged at high speed in cesium chloride salt.

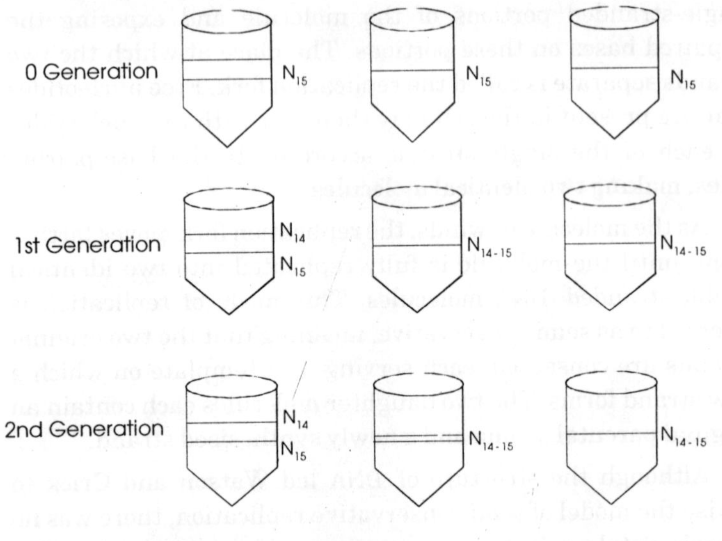

Figure 3. 5. Meselson-Staul experiment

When a concentrated solution of cesium chloride is placed under high centrifugal force, it generates a density gradient with the highest density at the bottom of the centrifuge tube and increasingly less dense matter toward the top. When DNA is included with the salt during centrifugation, it moves to an equilibrium point in the tube, at which the centrifugal forces moving it toward the bottom equal the buoyant forces formed by the density gradient pushing it toward the top. DNA containing ^{15}N is more dense than DNA containing ^{14}N and reaches equilibrium at a lower point in the tube than the DNA with ^{14}N. Using this method, Meselson and Stahl could distinguish the heavier ^{14}N DNA from the lighter ^{14} N DNA, as well as DNA containing half ^{15}N and half ^{14}N.

The three models for DNA replication predict where the DNA will equilibrate in the gradient. The conservative DNA replication model predicts that after one cell division, there will be two types of DNA that appear as separate fractions in the tube, one heavy (^{15}N) and one light (^{14}N).

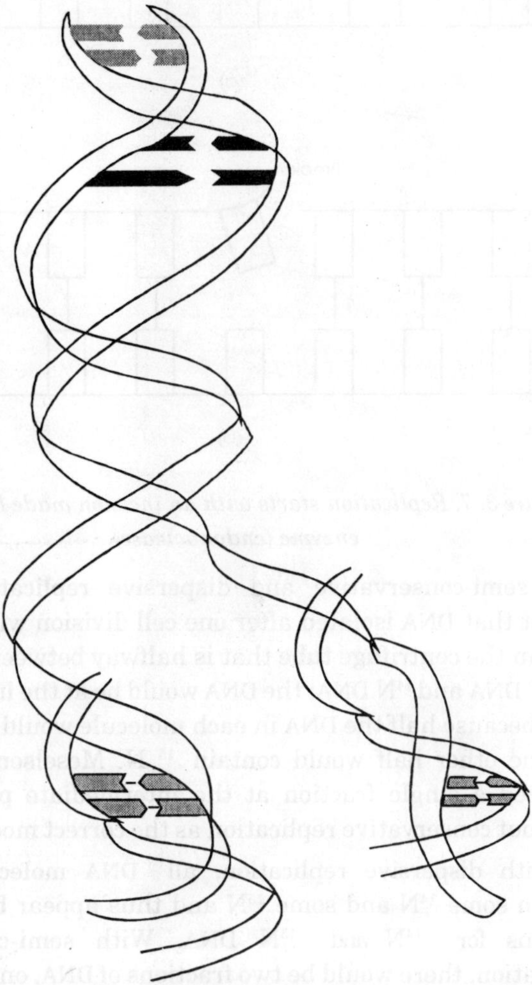

Figure 3. 6. Replication of DNA

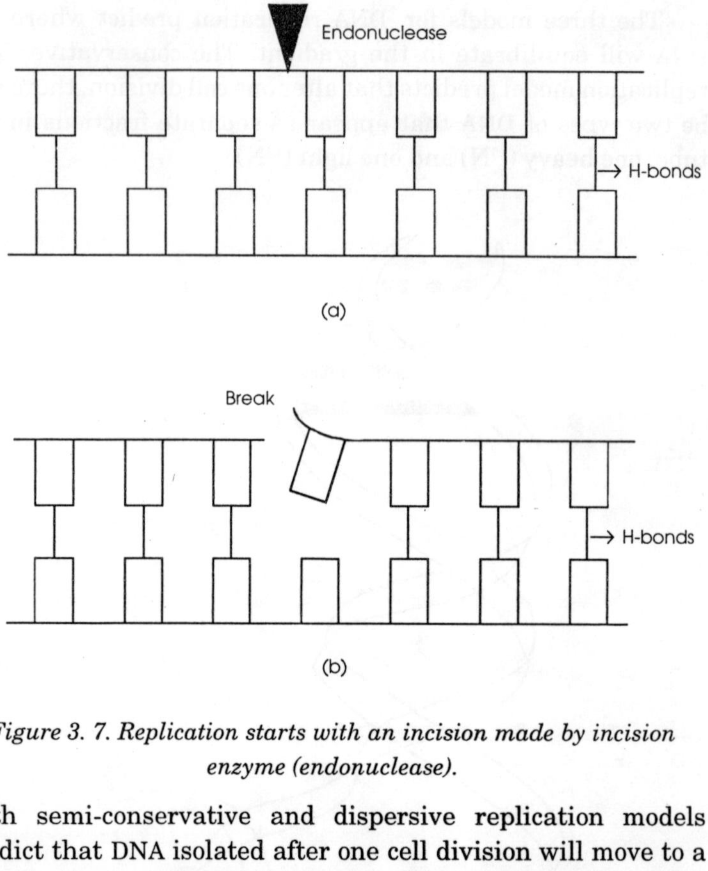

Figure 3. 7. Replication starts with an incision made by incision enzyme (endonuclease).

Both semi-conservative and dispersive replication models predict that DNA isolated after one cell division will move to a point in the centrifuge tube that is halfway between the points for ^{15}N DNA and ^{14}N DNA; the DNA would be at the intermediate point because half the DNA in each molecule would contain ^{14}N and the other half would contain ^{15}N. Meselson and Stahl observed a single fraction at the intermediate point, which ruled out conservative replication as the correct model.

With dispersive replication, all DNA molecules would contain some ^{14}N and some ^{15}N and thus appear between the positions for ^{14}N and ^{15}N DNA. With semi-conservative replication, there would be two fractions of DNA, one appearing at the point for ^{14}N DNA and the other at the intermediate point. Meselson and Stahl observed these two fractions in the DNA from cells that had divided twice, which eliminated the dispersive model and confirmed that DNA replicates semi-conservatively.

Separation and Stabilisation of DNA Strands

For DNA replication, the two DNA strands must separate into single strands. As the strands separate, the molecule must be allowed to unwind or excessive tension will build-up in it. As the molecule unwinds for replication, the part ahead of the replication fork may supercoil. The supercoiling is said to be positive because it turns in the right-handed direction, the same orientation as the original winding of the DNA. The tension of supercoiling must be released during DNA replication.

Enzymes called helicases unwind the DNA as the replication fork proceeds along the molecule, while enzymes called topoisomerases relieve the positive tension created by the unwinding. Although topoisomerase function is not fully understood, these enzymes apparently nick one strand of the double-stranded DNA ahead of the replication fork. The molecule then rotates at these nicks, relieving the tension of supercoiling. There is some indication that topoisomerases may also create negative supercoils ahead of the replication fork.

Negative supercoils turn in the direction opposite that of the original winding in the DNA, i.e., in a left-handed orientation. These negative supercoils may assist helicases because as the negative tension is relieved, this may help drive unwinding as the replication fork proceeds.

Single-stranded segments of DNA that result from the unwinding must be stabilised to prevent them from reattaching to one another. This is accomplished by single-strand binding proteins (SSBs). These proteins rapidly bind to the newly separated single-stranded DNA and maintain it in a single-stranded state until that portion of DNA is replicated. SSBs are referred to as cooperative tetramers because four protein subunits form a tetramer that binds to single-stranded DNA, and once an SSB tetramer is bound to single-stranded DNA, it facilitates the binding of an adjacent SSB tetramer. Until all of the single-stranded DNA is bound with SSBs, this process continues from one SSB to the next.

Synthesis of New DNA Strands

After stabilisation, the single strands of DNA may serve as templates upon which new strands are synthesised by enzymes referred to collectively as DNA polymerases. Several DNA polymerases are known, each with the capability of adding new nucleotides in a chain, using a single strand of DNA as a template.

In *E. coli*, DNA polymerase III, a complex enzyme consisting of seven different protein subunits, synthesises most of the DNA molecule. DNA polymerases can add new nucleotides only to the 3'-OH group on a nucleotide already paired with the template strand, which imposes two restrictions. First, DNA polymerases can only extend a chain of nucleotides in one direction. Since nucleotides are added only to the 3' end of a DNA strand, DNA synthesis is said to proceed only in the 5'- 3' direction. Second, DNA polymerases cannot initiate DNA synthesis on a single-stranded DNA template; they can only extend a chain of DNA from a nucleotide already paired to the template strand.

Unlike DNA polymerases, RNA polymerases are capable of adding ribonucleotides to a single-stranded DNA template that has no nucleotides paired with it. RNA is very similar to DNA in composition and structure, and its nucleotide subunits may pair with the nucleotides in DNA to form a double-stranded DNA-RNA hybrid molecule. For initiation of DNA replication, an RNA polymerase called primase synthesises a short segment of RNA (10 to 60 nucleotides) on the DNA template, forming a short RNA-DNA hybrid.

The segment of RNA is called a primer, and it initiates DNA synthesis. At one end of the primer is a terminal nucleotide with a 3'-OH group onto which DNA polymerases can add a new deoxyribonucleotide. Once the first deoxyribonucleotide is added, using the single-stranded DNA as a template, it exposes a 3'-OH group onto which another nucleotide may be added.

The DNA polymerase may then continue to add a nucleotide onto the 3'-OH group of each previous nucleotide, eventually creating a double-stranded DNA molecule. The free nucleotides that are incorporated into the DNA strands are in the triphosphate form—that is, they have three phosphate groups attached to the 5' carbon.

As each nucleotide is added to the growing molecule, the two terminal phosphate groups are cleaved from the nucleotide. Since the triphosphate form of a nucleotide contains more energy than the monophosphate form, removal of phosphates releases energy, driving DNA replication. DNA polymerases; synthesise new DNA strands using a single strand of DNA as a template. DNA polymerases add nucleotides only in the 5'-3' direction. Since DNA polymerases cannot initiate synthesis, an RNA primer is required for initiation.

Semi-discontinuous DNA Replication

Because DNA strands are antiparallel, only one strand may be synthesised continuously. In a replication fork, only one strand has its 3'-OH group facing the replication fork. On this strand, DNA polymerases may add new nucleotides to each previous one, synthesising DNA in the direction in which the replication fork is moving. As the replication fork proceeds along the molecule, DNA polymerases add new nucleotides continuously. The new strand of DNA that is synthesised continuously is called the leading strand. The situation on the other new strand is different.

Newly added nucleotides have the 3'-OH group facing away from the replication fork, meaning that DNA is synthesised in the direction opposite that in which the replication fork is moving. As the DNA molecule continues to unzip and the replication fork moves ahead, the newly exposed singlestranded DNA on the upper template strand cannot be used as a template for replication until a new primer is added, because the previous primer has a 5-phosphate group facing the replication fork.

Thus, the strand on the upper template must be synthesised discontinuously (i.e., in fragments) rather than continuously For this reason it is often called the lagging strand. The fragments of DNA formed by discontinuous DNA synthesis of the lagging strand are approximately 1, 000-2, 000 nucleotides long in prokaryotes and are much shorter (about 100-200 nucleoticles) in eukaryotes. These DNA fragments have been named Okazaki fragments after their discoverer, Reiji Okazaki.

Since one DNA strand is synthesised continuously and the other discontinuously, DNA replication is said to be semi-discontinuous. Synthesis of each fragment of the lagging strand requires a new initiation of DNA synthesis with a new RNA primer. How can an intact, double-stranded DNA molecule form when there are fragments, each beginning with an RNA primer? Somehow, the RNA primers must be removed and DNA must be synthesised in the space formerly occupied by the primers. Then, adjacent DNA fragments must be linked to one another.

In *E. coli*, DNA polymerase III terminates DNA synthesis when it reaches the primer of the previous fragment, leaving a nick between the 3' end of the newly synthesised fragment and the 5' end of the primer. At that point, the previous fragment's RNA primer must be removed and deoxyribonucleoticles must be added in its place. An enzyme called DNA polymerase I can do this because, in addition to its 5' → 3' polymerase activity, it also has a 5' → 3' exonuclease activity, meaning that it can remove nucleoticles ahead of it in the 5'→ 3' direction.

DNA polymerase I recognises the nick between the DNA fragment and the primer and binds at this site. It then migrates along the molecule, removing the primer ribonucleo-tides ahead of it. At the same time, it adds deoxyribonucleo-ticles to the template using its 5'→ 3' polymerase activity. After the final deoxyribonucleotide is added, a nick remains between its 3'-OH group and the 5'-phosphate group of the adjacent nucleotide. An enzyme called DNA ligase seals this nick by bonding the 3'-OH group to the 5'-phosphate group, creating a continuous doublestranded segment of DNA.

Synthesis of Leading and Lagging Strands

Models of DNA replication depict synthesis of the leading and lagging strands by separate DNA polymerase molecules. There is, however, evidence that the two strands are synthesised simultaneously. In *E. coli*, DNA polymerase III is a large enzyme composed of many protein subunits. The subunits that replicate DNA are designated α. Noting that there are two α subunits, investigators hypothesised that one subunit might synthesise the leading strand while the other synthesises the lagging strand.

The β subunits assemble to form guides through which double-stranded DNA can pass. So how can a single protein complex simultaneously synthesise two DNA strands that are antiparallel? If the template for the lagging-strand forms a loop, then the looped portion points in the same direction as the template for the leading strand. Then, the DNA polymerase III holoenzyme can simultaneously synthesise both strands in the same direction, one continuously and the other discontinuously.

Proof-reading of Newly Synthesised DNA

Proofreading by DNA polymerases corrects most errors in DNA replication. If prokaryotic DNA replication relied entirely on chemical interactions between nitrogenous bases for accurate base pairing during the synthesis of a new strand of DNA, bases would be mispaired about once every 1,000 nucleotides. However, actual errors during DNA replication are far fewer, about once every billion or more nucleotides. This remarkable fidelity in DNA replication is due mostly to proofreading by DNA replication enzymes.

In addition to their 5'→3' polymerase activity, prokaryotic DNA polymerases also have a 3'→ 5' exonuclease activity. This means that these polymerases are capable of moving backward and removing nucleotides from DNA in the direction opposite that of DNA synthesis. When the DNA polymerase adds an incorrect nucleotide during DNA synthesis, the DNA polymerase apparently recognises the error and uses its 5'→3' exonuclease activity to remove the mismatched nucleotide. It then adds the correct nucleotide in its place and continues DNA synthesis in the 5'- 3' direction.

Nucleotides may be mismatched and DNA may be damaged in a number of ways. If left uncorrected, the errors may become mutations that alter the sequence of nucleotides. The main function of the *E. coli* enzyme DNA polymerase H is to synthesise DNA during repair of DNA damage. There are many mechanisms that cells may use to correct nucleotide mismatches or repair DNA damage. Although highly effective, DNA proofreading and repair mechanisms are not perfect, so occasionally DNA mutates and the mutations persist during subsequent DNA replication.

Replication of Eukaryotic DNA

The mechanism for DNA replication in eukaryotes is similar to that in prokaryotes, but the enzymes differ. Most of the enzymes responsible for DNA replication in yeast and in mammals have now been characterised. There are five known DNA polymerases in mammals, named α, β, δ, ε and γ. DNA polymerases α and δ are both essential for replication and are probably the DNA polymerases responsible for most of the DNA synthesis.

There is strong evidence that α is responsible for synthesis of the lagging strand, and δ is responsible for synthesis of the leading strand. DNA polymerase α also functions as a primase, synthesising RNA primers on a DNA template for initiation of replication. The two smallest subunits in association may act as a primase on their own. They may also act as a primase when they are part of DNA polymerase α. The 180 and 70 kD subunits of DNA polymerase a function in DNA replication. The combined DNA polymerase and primase activities of DNA polymerase α constitute strong evidence that this enzyme synthesises the Okazaki fragments that become the lagging strand.

DNA polymerases ε and β help repair damaged DNA and associate with exonucleases to remove damaged DNA. The fifth DNA polymerase in eukaryotes, γ, is found only in mitochondria. This enzyme synthesises and proofreads the organelles DNA. DNA polymerases δ, ε, and γ have 3'→ 5' exonuclease activities, indicating that they are capable of proofreading. None of the eukaryotic DNA polymerases have a 5'→ 3' exonuclease activity like that of DNA polymerase I in *E. coli*. Instead, there are several separate enzymes in eukaryotes that remove nucleotides in the 5'→ 3' direction.

REPLICATION OF ENTIRE DNA MOLECULES

DNA molecules may be circular or linear, and they range in size from a few thousand nucleotide pairs in viral DNAs to more than a hundred million nucleotide pairs in some eukaryotic chromosomes. Cells and viruses employ several different strategies for replicating their DNA molecules. Which strategy is best depends on the molecule's size and conformation, and on specific nucleotide sequences that determine how the molecule will replicate.

Origins of Replication

Origins of replication are fixed sites in DNA that separate easily because they are rich in A-T pairs. Replication does not begin at random along the molecule, but rather at predetermined positions that are used each time the DNA replicates. The nucleotide sequences at some origins of replication have been identified, and they tend to be rich in A-T base pairs. Since each A-T pair is held together by only two hydrogen bonds, whereas a G-C pair is held together by three, the strands in a segment of DNA containing mostly A-T pairs are more easily separated than those in a segment containing many G-C pairs.

The single origin of replication in *E. coli* is called as oriC. The oriC origin of replication consists of 245 nucleotide pairs including three segments having the same sequence of 13 nucleotide pairs, called a 13-mer, followed by four segments having the same sequence of 9 nucleotide pairs (a 9-mer). The 9-mers, direct a set of proteins, called DnaA proteins, to bind to the DNA in the region of the 9-mers. The DnaA protein binding causes the DNA to coil into a loop, which, in turn, causes the DNA in the region of the 13-mers to unwind, allowing replication to begin.

Bidirectional Replication

A replication fork could begin at an origin of replication on one end of a linear DNA molecule and proceed in a single direction throughout the molecule until the entire molecule was replicated. Or, in a circular molecule, a replication fork could begin at a single origin of replication in the circle and proceed around in one direction until the entire circle was replicated. Replication that proceeds in this way is known as unidirectional replication. DNA replication is unidirectional in some viruses and bacterial plasmids.

However, in most cases DNA replication begins at an origin of replication somewhere within the molecule, and then two replication forks begin in opposite directions from the origin, a phenomenon known as bidirectional replication. Most eukaryotic DNA molecules are linear and typically orders of magnitude longer than bacterial DNA molecules. Each of these long molecules has many origins of replication that can initiate

DNA replication simultaneously, enabling the DNA to replicate more quickly than it could from a single origin of replication. As replication forks proceed in both directions from the origins, they form bubbles of replicated DNA that eventually merge until the entire molecule is replicated.

Replication of Circular DNA

Nearly in every cell circular DNA molecules can be seen. The large DNA molecules of bacterial cells are circular, as are the smaller plasmid molecules. In eukaryotic cells, mitochondria contain circular molecules, as do chloroplasts in plant cells. There are three basic strategies for replicating circular molecules: (i) θ–mode repication, (ii) σ-mode or rolling circular replication and (iii) D-loop replication.

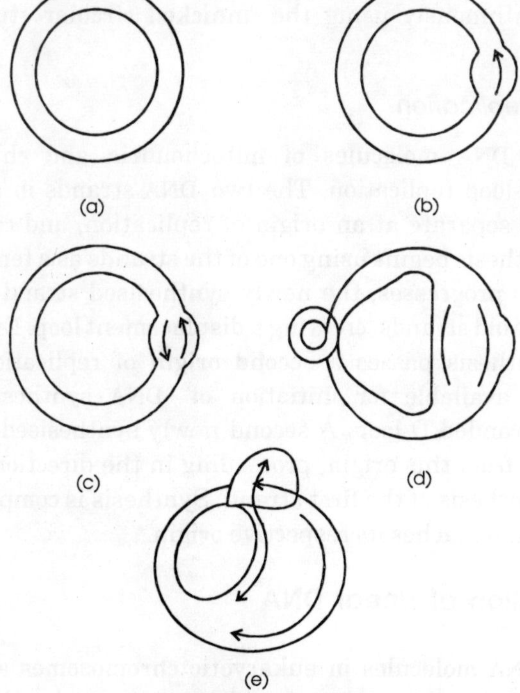

(a) (b) (c) (d) (e)

Figure 3.8. Replication of circular DNA

θ-mode replication

Most large circular DNA molecules in bacteria have a single origin of replication from which replication forks proceed bidirectionally to form a q-shaped intermediate when replication is partially completed. Eventually, the two replication forks meet and the two circular molecules separate.

s-mode or rolling circle replication

This strategy of DNA replication has a σ-shaped intermediate. Several bacterial viruses use σ-mode replication to rapidly make multiple copies of the viral DNA. A nick is made at the origin of replication in one strand of the double-stranded DNA, and the nicked strand separates from its complementary strand beginning at the 5' end of the nicked site. The nicked strand is used as a template for discontinuous DNA synthesis forming Okazaki fragments, while a second strand of DNA is synthesised continuously using the unnicked circular strand as a template.

D-loop replication

Circular DNA molecules of mitochondria and chloroplasts utilise D-loop replication. The two DNA strands in a circular molecule separate at an origin of replication, and continuous DNA synthesis begins using one of the strands as a template. As synthesis progresses, the newly synthesised strand displaces one of the old strands, creating a displacement loop. Eventually, DNA synthesis passes a second origin of replication, which becomes available for initiation of DNA synthesis in the single-stranded D-loop. A second newly synthesised strand is initiated from this origin, proceeding in the direction opposite to the synthesis of the first strand. Synthesis is complete when each strand reaches its respective origin.

Replication of Linear DNA

Large DNA molecules in eukaryotic chromosomes are linear. What we have learned so far about DNA replication cannot explain how the ends of linear molecules can be fully replicated. To understand why the ends of DNA molecules cannot be fully replicated using semi-disconfinuous replication, think about primers. If a primer forms at the end of a linear DNA molecule to initiate replication of the leading strand, a single-stranded gap remains at that end after DNA has been synthesised and the primer has been removed. DNA polymerases cannot fill the gap because there is no 3'-OH group onto which nucleotides can be added. If this gap were left unfilled, a linear DNA molecule would become shorter at each end by the length of the primer with each replication.

After replicating many times, the molecule would eventually disappear. This problem is called the linear DNA replication paradox. If cells fail to fully replicate the ends of their DNA molecules, eventually the molecules erode at the ends, losing important information, which halts cell growth and division. For cells to grow and divide for more than a few generations, they must have a strategy for overcoming the linear DNA replication paradox. Linear DNA molecules found in nature employ several different strategies for overcoming the linear DNA replication paradox. In perhaps the simplest strategy, a linear DNA molecule arranges itself into a circle in preparation for replication.

REFERENCES

Avery, O. T., MacLeod, C. M. and McCarty, M., 'Studies on the Chemical Nature of the Substance Inducing Transformation of the Pneumococcal Types', *Journal of Experimental Medicine*, 79:137-158.

Campbell, J. L., 'Eukaryotic DNA Replication', *Annual Review of Biochemistry,* 55:733-771, 1986.

Crick, F., *What Mad Pursuit: A Personal View of Scientific Discovery*. New York: Basic Books, 1988.

Franklin, R. E., and R. G. Gosling, 'Molecular Configuration in Sodium Thymonucleate', *Nature,* 171:740-741, 1953.

Greider, C. W., and Blackburn, E. H., 'Telomeres, Telomerase, and Cancer', *Scientific American*, 267 (Feb): 92-97, 1996.

Griffith, F., 'The significance of pneumococcal types', *Journal of Hygiene*, 27:141-144, 1928.

Hershey, A. D., and Chase, M., 'Independent Functions of Viral Protein and Nucleic Acid in Growth of Bacteriophage', *Journal of General Physiology,* 36:39-56, 1952.

Judson, H. E., *The Eighth Day of Creation*, New York: Simon & Schuster, 1979.

Kornberg and T. A. Baker, *DNA Replication*, 2nd ed. New York: Freeman, 1992.

Meselson, M., and Stahl, F. W., 'The Replication of DNA in *Escherichia coli*'. *Proceedings of the National Academy of Sciences*, USA 44:671-682, 1958.

Mullis, K. B., 'The Unusual Origin of the Polymerase Chain Reaction', *Scientific American* 262 (Apr): 56-65, 1990.

Sambrook, J., Fritsch, E. E and Maniatis, T., *Molecular Cloning: A Laboratory Manual*, 2nd ed. Cold Spring Harbor, N.Y.: Cold Spring Harbor Laboratory Press, 1989.

4

Chromosome Mapping

According to Mendalian genetics, genes were known to be located on chromosomes, and independent assortment of chromosomes explained independent assortment of genes. However, there were many more genes than chromosomes, so some genes had to be located on the same chromosome. These genes should remain inextricably associated with one another, failing to recombine through independent assortment.

Hugo de Vries attempted to resolve this dilemma by proposing that homologous chromosomes freely exchange genes with one another when the chromosomes pair during meiosis. If this is true, all genes on the same chromosome should assort independently, just like genes on non-homologous chromosomes.

Although de Vries's proposal was incorrect, he was not far from the truth. Through crossing-over, paired homologous chromosomes exchange segments, but the exchanges are not frequent enough for all genes to assort independently. Genes located on the same chromosome are called syntenic genes. Some syntenic genes are so distant that they assort independently during meiosis because crossing-over between them is frequent. On the other hand, some syntenic genes are linked genes: they are so close to one another that crossovers may recombine them, but the crossovers are not frequent enough to permit independent assortment.

Frequency of recombination between two genes depends on the distance between them on the chromosome. The closer two genes are, the lower the frequency of crossovers between them.

The farther apart they are, the greater the frequency of crossovers. According to some geneticists, crossovers are random events along the length of the chromosome, which explains why distantly linked genes are more likely to have a crossover between them than closely linked genes.

Actually, crossing-over is not entirely random; it is less frequent in certain regions of the chromosome, such as the region surrounding the centromere. But the general rule that crossing-over is more frequent between distant genes than closely linked genes still holds true. Thus, scientists can measure the frequency of recombination between linked genes to determine the relative positions of genes on a chromosome, a process called chromosome mapping.

LINKAGE AND CROSSING-OVER

Linkaging the nature of genes was first reported in 1900, by Carl Correns. Correns reported a case of complete linkage, a situation in which two genes remain completely associated with each other when transmitted from parents to offspring. Correns hybridised two types of stocks, one with purple flowers and hoary leaves, the other with white flowers and smooth leaves. The F_1 plants had purple flowers and hoary leaves, indicating that these two traits were dominant. In the F_2 progeny, Correns expected the two traits to assort independently. However, all purple-flowered F_2 progeny had hoary leaves, and all white-flowered F_2progeny had smooth leaves, just like the two original parents. These two phenotypes appeared in a 3:1 ratio. Such an observation could be attributed to pleiotropy.

Recall Mendel's experiments in which flower colour and seed coat colour were associated with one another because the two traits were governed by the same gene, an example of pleiotropy. In Correns's experiments, the purple flowers in stocks were always associated with coloured seeds, which Correns correctly interpreted as pleiotropy. However, Correns knew of stocks with purple flowers and smooth leaves as well as plants with white flowers and hoary leaves, an indication that the two traits are not fully dependent on one another. Correns correctly concluded that flower colour and leaf texture were governed by different genes that fail to assort independently.

Incomplete linkage of gene was reported by William Bateson and Reginald Punnett in 1908 in sweet pea. They observed that in the F_2 progeny of a cross between a purple-flowered plant with long pollen grains and a red-flowered plant with round pollen grains, these traits failed to assort independently, although all four possible phenotypes were present. However, linkage analysis in F_2 progeny can be a complicated situation and Bateson and Punnett initially developed an incorrect theory to explain the deviation from independent assortment.

It was Thomas Hunt Morgan who correctly identified linkage. He and his co-workers hybridised flies with several mutant alleles at different X-linked loci and discovered that the genes did not assort independently among the offspring. Morgan concluded that the deviations from independent assortment were caused by linkage. Other scientists had reported the appearance of chiasmata in prophase I chromosomes. Morgan assumed that they represented exchanges of chromosomal segments, which he called crossovers, and that crossovers were responsible for recombination of syntenic genes.

Morgan's student, Alfred Sturtevant, later compiled data that he and others had collected and used these data to determine the relative positions of five genes on the X chromosome of *Drosophila.* Although the title of Sturtevant's paper "The linear arrangement of six sex-linked factors in *Drosophila,* as shown by their mode of association" suggests that he mapped six genes, he actually mapped only five loci. Two of the genes he studied, *white* and *eosin,* were both eye colour alleles and showed complete linkage. It was later discovered that they were different alleles of the *white (w) locus.* Sturtevant also studied mutant alleles at the *y* (yellow body), v (vermilion eyes), *r* (rudimentary wings), and *m* (miniature wings) loci, all on the X chromosome.

After analysing data from 31 different crosses, Sturtevant created the chromosome map. The numbers show the frequencies of recombination between alleles of the *w*, *v*, *r*, and *m* loci with alleles of the *y* locus, to which Sturtevant assigned a position numbered zero. This was the first chromosome map ever made. Sturtevant, together with Calvin Bridges, later discovered linkage on autosomes in *Drosophila*. Eventually, detailed genetic maps for several model organisms were made from the results of experiments conducted over a period of several decades. The units used for mapping genes on chromosomes represent the frequency of crossovers as a percentage value and are often called centimorgans (cM) in honour of Morgan's pioneering work.

Mapping is one of the most important areas of current genetics research. Human genome mapping is under way and, as we will see later in this chapter, mapping genes and DNA markers in a large number of species, including humans, is one of the most active areas in genetic research. Genetic map distances between linked genes are measured in centimorgans (cM), which are defined as the percentage of crossing-over between linked genes.

TWO-FACTOR LINKAGE ANALYSIS

Genetic map distances between two genes are typically calculated by conducting a two-factor testcross and dividing the number of recombinant progeny by the total number of progeny. Two linked genes may be in either coupling or repulsion conformation.

Typically, two-factor linkage is diagrammed using linkage notation, which consists of a line representing the chromosome, and the alleles of two linked genes designated by letters:

A *B*
____

The homologous chromosome is represented by a second line with the alleles on it. For instance, when loci *A* and *B* are linked, the cross *AA BB* X *aa bb* produces the F_1 heterozygote *Aa Bb*. In linkage notation, this cross is shown as follows:

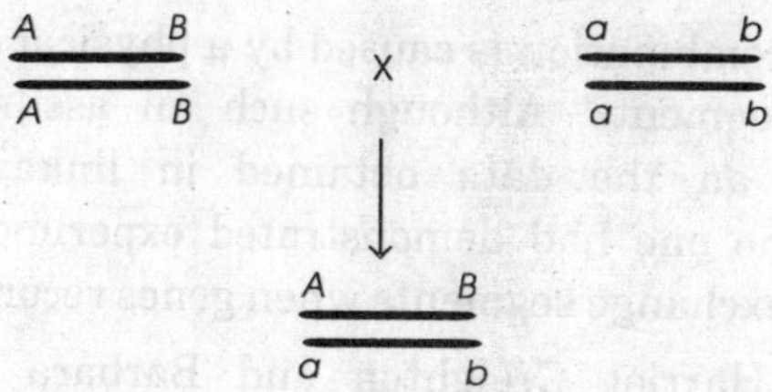

Crossovers between two genes can be detected in the progeny of an individual that is heterozygous for two linked genes. The chromosomes that arise from a crossover between two genes are called recombinant chromosomes.

In recombinant chromosomes, the alleles at two linked loci have recombined when compared to the original chromosomes. When there is no crossover between two linked genes, the chromosomes that arise are called non-recombinant chromosomes. The recombination frequency between two genes is calculated by dividing the number of recombinant chromosomes by the total number of chromosomes.

Linkage analysis is usually done with testcross progeny because each of the four phenotypes in testcross progeny represents one of the four possible genotypes. In testcross progeny, researchers can readily distinguish recombinant from non-recombinant chromosomes.

When analysing data from a dihybrid testcross, we must first ask if there is evidence to indicate that the two genes are linked. Since linkage is manifest as a deviation from independent assortment, we first determine whether or not the deviation is significant using a chi-square test. If the deviation is not significant, there is insufficient evidence to conclude that linkage is present. If the deviation is significant, then linkage may be the most probable explanation for the deviation.

Physical Detection of Crossovers

Recombination of linked genes is associated with cytological observation of crossing-over. Crossing-over is initiated between non-sister chromatids after chromosomes have been duplicated.

Genetic recombination is caused by a physical exchange of chromosome segments. Although such an assumption was supported by an the data obtained in linkage analysis experiments, no one had demonstrated experimentally that chromosomes exchange segments when genes recombine.

In 1931, Harriet Creighton and Barbara McClintock reported the results of what is now considered a classic experiment to determine whether genetic combination of linked gene is associated with a physical exchange of chromosomal segments. They used two genes the *c*1 locus and the *wx* locus on chromosome 9 in maize, from plants that had cytological markers, differences in the chromosomes that can be observed microscopically.

Creighton and McClintock found a chromosome with 9 with two cytological markers, a knob on one end and a piece from another chromosome on the other end. Thus, both ends of this chromosome could be distinguished from normal copies of chromosome using microscopy. This chromosome carried the dominant *C*1 allele and recessive *wx* allele. From the cross they observed the following progeny:

*c*1*c*1 *Wx wx*	knobless	no translocation	5
*c*1*c*1*Wx Wx*	knobless	no translocation	5
*c*1*c*1 *Wx* —	knobless	no translocation	1
*C*1*c*1 *wx wx*	knob	translocation	3
*c*1*c*1 *Wx Wx*	knobless	translocation	2
*c*1*c*1 *Wx wx*	knobless	translocation	2
*c*1*c*1 *wx wx*	knobless	translocation	2
*C*1*c*1 *Wx*—	knob	no translocation	2

Crossing-over in Duplicated Chromosomes

Chromosomes initiate crossing-over during prophase 1 of meiosis after they have duplicated, so a homologous chromosome pair contains four chromatids: two sister chromatids in one chromosome, and two sister chromatids in its paired homologue. Crossing-over between two sister chromatids, also known as meiotic sister chromatid exchange, is probably rare in meiosis and has no genetic consequence. Sister chromatids are identical to one another, so a crossover between two sister chromatids generates identical chromatids and has no effect on inheritance. Non-sister chromatids on paired homologous chromosomes of a doubly heterozygous individual must exchange segments, for crossovers to be detected. It does not matter which non-sister chromatids exchange segments.

MAP DISTANCE CORRECTION

Any factors that alter crossover frequencies need to be taken into account in order to calculate accurate genetic map distances. The chromosomal location of a gene, the age, sex, or genotype of an individual, and environmental influences may all affect crossover frequency. In some cases, these effects may be negligible, but in other cases they may be significant. For example, in *Drosophila melanogaster,* there are no crossovers in male flies, and as female flies grow older, the frequency of crossovers decreases. Temperatures either warmer or cooler than 22°C tend to increase the frequency of crossing over *Drosophila* females, and the levels of calcium and magnesium in the diet also influence crossover frequencies.

Researchers can compensate for these factors by designing crosses so that the female is always the heterozygous parent in a testcross, by using young female parents in mapping crosses, and by controlling the environment in which the flies are raised. Factors that alter crossover frequency tend to differ among species, so researchers need to be aware of the peculiarities for the species in question. Generally, experiments can be designed to overcome these alterations, or corrections can be made if the nature of the alteration is known beforehand. For example, in humans crossover frequencies differ in males and females.

Map distances are lower when derived from male parents than from female parents, an observation that implies a lower crossover frequency in males. For this reason, human geneticists must adjust map distances to compensate for the differences in crossover frequency between males and females.

Mapping Function

Two-factor mapping experiments tend to underestimate the actual map distance when the true genetic map distance exceeds 20 cM. For example, suppose that genes *A* and *B* map at 13 cM from each other, and that genes *B* and C map at 18 cM from each other in separate experiments. *A* and C should map at 31 cM from each other based on the additive distances of *A-B* and *B-C*. However, the experimental results from the progeny of a two-factor cross for *A* and C consistently yield a map distance less than 31 cM.

Underestimation of map distances is the result of undetected double crossovers. A double crossover is the simultaneous occurrence of two crossovers between linked genes. Chromosomes exchange segments during prophase I when paired chromosomes are already duplicated. This means that any two non-sister chromatids can participate in the crossover.

There are four possible types of double crossovers, some of which are detected phenotypically, but some of which are not.

A single crossover produces two recombinant and two non-recombinant chromosomes. Thus, for every single crossover, two of four chromosomes are recombinant after meiosis is completed.

A two-strand double crossover produces four non-recombinant chromosomes, even though there are two crossovers. Phenotypically, this appears as though there were no crossovers.

Three-strand double crossovers produce two recombinant chromosomes and two non-recombinant chromosomes. Phenotypically, these appear as a single crossover.

A four-strand double crossover produces four recombinant chromosomes. Phenotypically, this type of crossover produces the same number of recombinant chromosomes per crossover as single crossovers do.

In a two-factor cross, two-strand and three-strand double crossovers result in undetected crossovers. The frequencies of the four types of double crossovers are equal. Thus, half of the crossovers from double crossovers are not detected phenotypically.

Expected frequency of double crossovers between two genes is the square of the genetic map distance between the two genes. This means that the probability of double crossovers should increase exponentially as the distance between genes increases. When genetic map distances are very small, the frequency of double crossovers is so small that the probability of undetected crossovers may be negligible.

The presence of one crossover interferes with initiation of a second crossover in its immediate vicinity a phenomenon known as interference. When the distances between genes are small, the probability of a double crossover between them is close to zero. At increasing distances between genes, the probability of double crossovers increases exponentially, meaning that at some point the number of undetected crossovers begins to significantly affect the calculation of genetic map distances. The frequency of crossing-over has a curvilinear relationship to the phenotypic recombination frequency. Phenotypic recombination frequencies can be corrected to approximate crossover frequencies with equations called mapping functions.

The first mapping function, often called the Haldane function, was derived by J. B. S. Haldane in 1919, and is expressed as

$$x = -0.5\ln(1 - 2y) \quad (1)$$

Where, x is an estimate of the actual frequency of crossing-over, ln stands for natural logarithm, and y is the phenotypic recombination frequency. This is a theoretical equation and does not take interference into account, so it departs from the linear mapping function at zero.

In reality, interference causes the mapping function to remain almost linear between 0 and 20 cM. Haldane recognised this in his original paper and developed a second function based on empirical data instead of theory. Since Haldane's attempt to derive a mapping function to match empirical data, several researchers have derived mapping functions from empirical data for particular species. Perhaps the most widely used is the Kosambi function. This function, derived in 1944 by D. D. Kosambi, can be used to correct phenotypic recombination frequencies measured in *Drosophila* and many other species.

As in equation, x estimates the frequency of crossing-over and y equals the phenotypic recombination frequency:

$$x = 0.25\ln\frac{1 + 2y}{1 - 2y}$$

Theoretical Haldane function tends to overestimate the true genetic map distance when interference is present; functions derived from empirical data are usually better because they take interference into account.

Genetic map distance in F_2 progeny

Some genotypes carry two recombinant chromosomes, some carry one recombinant and one non-recombinant chromosome, and others carry two non-recombinant chromosomes, all of which may be found in a single phenotypic class. For this reason, calculation of map distances from phenotypic frequencies is difficult and not highly reliable statistically, so whenever possible, it is best to do genetic mapping in testcross progeny. Phenotypically, linkage in F_2 progeny is manifest as a deviation from the 9:33:1 ratio predicted for independent assortment.

THREE-FACTOR LINKAGE ANALYSIS

It is possible to map three or more linked genes in the progeny of a single cross. In a testcross situation the map distance between two genes is calculated by counting the number of progeny that arose from a crossover between the two genes and dividing that number by the total number of progeny. In a three-factor testcross, this procedure is done twice.

A B C

With three linked genes the genetic map distance between *A* and *B* and the genetic map distance between *B* and *C* are calculated separately using progeny from the same cross. However, before calculating the map distances, we must determine the order of the three genes. With three genes, *A, B,* and *C*, there are three possible orders:

A B C

B A C

A C B

We usually write the order of genes on a chromosome from left to right, because that is the direction we read. But in a three-factor testcross, we determine the order of three genes in relation to one another. Under these circumstances, the left-right orientation does not matter; A B is the same as C B A. We determine the correct order from the proportions of individuals in each class of testcross progeny.

The three-factor testcross *Aa Bb Cc* X *aa bb cc* produces eight genotypic and phenotypic classes. If the alleles at the three loci assort independently, the eight genotypes should appear in a 1:1:1:1:1:1:1:1 ratio. Linkage causes a significant deviation from this ratio. Let's suppose for the time being that the genes are linked and that their order is *A B C*, so the heterozygous parent has the genotype *ABC/a b c*.

Of the eight possible gametes that could arise from the heterozygous parent, two gametes arise from no crossovers in the region of interest *(A B C* and *a b* c), two from a single crossover between *A* and *B (A b c* and *a B C*), two from a single crossover between *B* and C *(A B* c and *a b* C), and two from a double crossover (one crossover between *A* and *B* and a second crossover between *B* and C: *A b* C and *a B c*).

The most frequent chromosomes are the non-crossover types. Using this information, we can assign each class of progeny to its crossover type. The double-crossover classes are the least frequent, and the non-crossover types are the most frequent. To determine the gene order on the chromosome, we diagram the three possible orders and determine which produces the observed double-crossover class. If the gene order on the chromosome is *A B C,* then the double-crossover classes are:

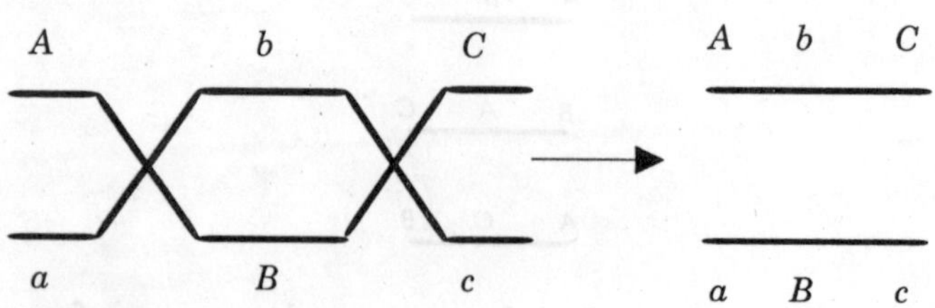

If the gene order on the chromosome is *B A* C, then the double-crossover classes are:

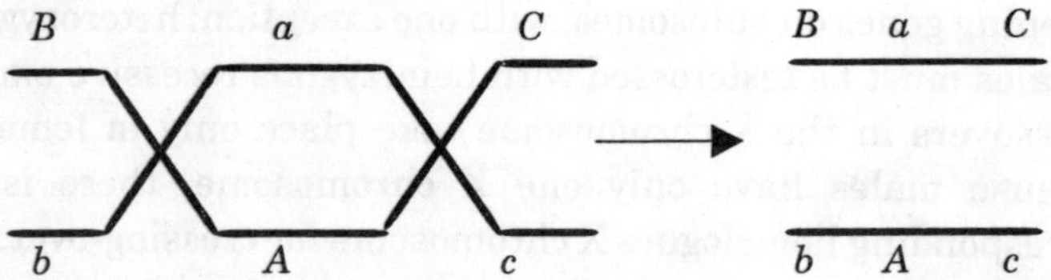

And if the gene order on the chromosome is *A C B,* then the double-crossover classes are:

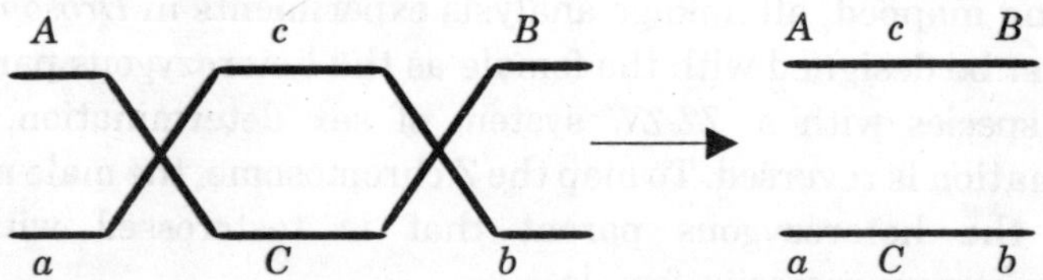

After determining gene order, genetic map distance is calculated. Let's assume that the gene order is *A B C*. As with two-factor testcross progeny, we calculate the genetic map distance between A and *B* by determining the number of progeny that represent a crossover between *A* and *B* and dividing this number by the total number of progeny. This means that we add the totals for progeny that represent a single crossover between *A* and *B* and all double-crossover progeny, then divide this sum by the total number of progeny. To calculate the map distance between *B* and *C*, we add the totals for progeny that represent a single crossover between *B* and C and all double-crossover progeny, then divide this sum by the total number of progeny.

Mapping Genes on Heteromorphic Chromosomes

Sex-determining chromosomes such as X and Y chromosomes or Z and W chromosomes are heteromorphic. Mapping genes on the X chromosome in a species with an XX-XY system of sex determination requires essentially the same procedure as mapping genes on autosomes, with one exception: heterozygous females must be testcrossed with hemizygous recessive mates. Crossovers in the X chromosome take place only in females; because males have only one X chromosome, there is no corresponding homologues X chromosome for crossing-over.

For mapping testcrosses, the XX females must be heterozygous for the loci being mapped, and they are testcrossed with recessive XO males. In *Drosophila,* there is no crossing-over in the male, so regardless of the chromosome being mapped, all linkage analysis experiments in *Drosophila* must be designed with the female as the heterozygous parent. In species with a ZZ-ZW system of sex determination, the situation is reversed. To map the Z chromosome, the male must be the heterozygous parent that is testcrossed with a hemizygous recessive female.

MULTIFACTOR CHROMOSOME MAPPING

We can calculate map distances with four or more linked genes by extending the procedures used for three-factor mapping. With each added factor, the number of possible genotypes and phenotypes in testcross progeny doubles. This tends to make mapping beyond a three-factor cross rather cumbersome when done by hand. For example, there are 16 possible genotypes with four-factor linkage. In four-factor linkage, the coupling-repulsion relationships of four genes must be determined and non-crossover, single, double, and triple-crossover-type progeny must be identified.

Multifactor chromosome mapping is typically done using computer programmes that construct genetic maps from data sets. With DNA marker analysis, it is possible to construct a genetic map of all chromosomes from the progeny of a single cross. Genetic markers may include alleles that confer outward phenotypic differences or DNA markers. Genetic markers are initially mapped to linkage groups. A linkage group is a chromosomal segment that contains two or more linked genetic markers. As large numbers of genetic markers are mapped in a species, linkage groups increase in size and may be coalesced as markers from one linkage group are found to be linked to markers in another linkage group. Eventually, the number of linkage groups equals the haploid number of chromosomes, and each linkage group represents one chromosome.

For traditional gene mapping, at least two polymorphic markers must be present at a locus. Polymorphic markers are genetic markers that are distinguishable as different alleles of a single locus. The word polymorphic means literally of many forms. Polymorphic markers may be alleles that differ in some identifiable respect from each other. For example, the alleles that caused purple and white flowers in Mendel's peas are polymorphic to each other since they cause an identifiable difference.

DNA markers, such as RFLPs or RAPDs, are also polymorphic. Polymorphic DNA markers are homologous segments of DNA located at the same position on homologous chromosomes that can be distinguished from one another because of differences in DNA sequence. As the chromosomes on which these segments of DNA reside are inherited in a Mendelian fashion, polymorphic markers are likewise inherited in a Mendelian fashion, and linked DNA markers behave genetically just like linked genes. Linkage analysis with DNA markers is no different than analysis with traditional genetic markers.

There are some distinct advantages to DNA markers, however, that allow rapid chromosome mapping. The main advantage is the large number of potential markers available. In many species, a genetic map of all chromosomes may be made in the progeny of a single cross using DNA markers; many different crosses are typically required for traditional chromosome mapping. When DNA markers are used, a map can be constructed quickly by identifying DNA markers that are polymorphic between the two parents and mapping those markers in the progeny.

With traditional morphological markers, this is not possible in a single cross. With maize, for example, one cannot expect two single plants to differ in kernel shape, colour, and texture; leaf shape, pattern, texture, and colour; plant height and growth habit; male fertility; disease resistance; etc., without going through years of effort to breed differences for these traits into single plants. However, the number of DNA markers that can be tested for polymorphism numbers in the thousands, so it is often possible to identify large numbers of polymorphic DNA markers in two parents and map all of them in the progeny of a single cross.

GENETIC MAPPING IN HUMANS

J.H Tijo and A. Leven in 1956 determined for the first time, the correct chromosome numbers in human beings, i.e., 2n = 46. Later, on the basis of morphology, twenty three chromosomes in human beings were arranged in seven groups.

The numbers of offspring tend to be small compared to other species used for genetic analysis, and human crosses cannot be designed to meet experimental needs. Even with these limitations, however, statistical methods for use in human genetic mapping were developed in the 1950s. Unfortunately, only limited mapping could be done because of an insufficient number of polymorphic genetic markers. With extensive efforts, some genes were mapped on the X chromosome, and a few autosomal linkages were found.

Several methods for physical mapping of human chromosomes emerged and were used to establish the physical locations of some genes on their respective chromosomes. While extensive genetic maps of model organisms, such as *Drosophila,* mouse, and maize, were being developed over a period of decades, human genetic mapping remained almost stagnant. All of this changed when DNA marker analysis became available. Tens of thousands of polymorphic tandem repeat markers, such as minisatellites and microsatellites, can be detected in humans.

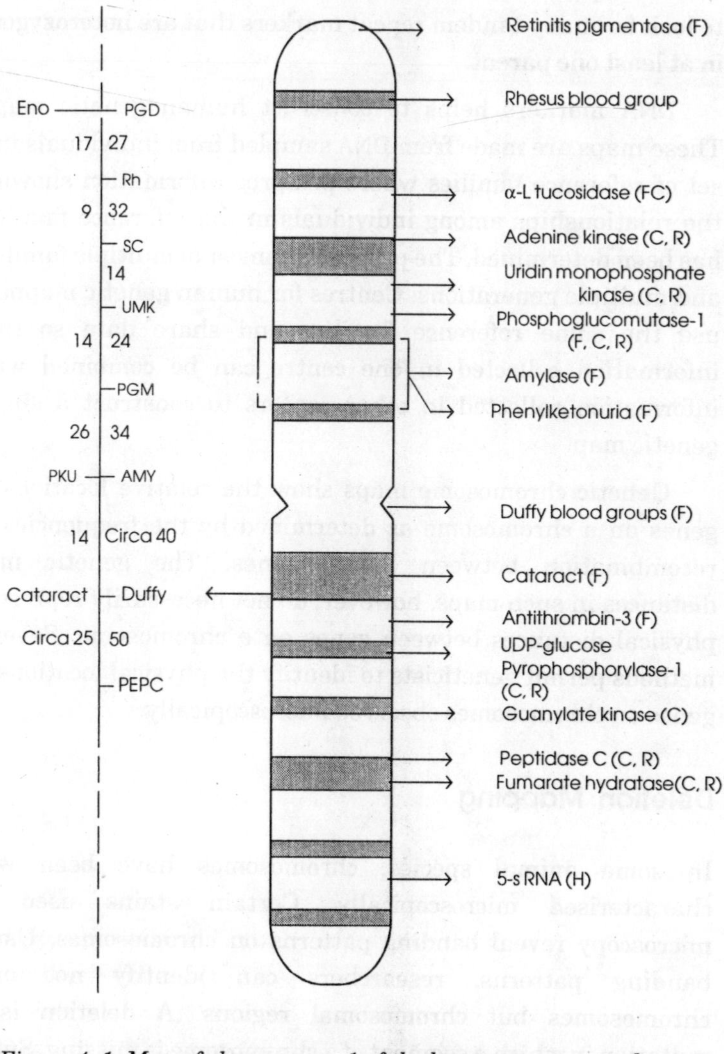

Figure 4. 1. Map of chromosome 1 of the human genome and genes that have been identified.

Most tandem repeat markers are codominant, so homozygotes and heterozygotes can be readily distinguished from one another. In order for a pedigree to provide information that can be used in mapping, at least one parent must be doubly heterozygous for linked markers. Since large numbers of tandem repeat markers are available, researchers simply need to look for many tandem repeat markers that are heterozygous in at least one parent.

DNA markers helps to construct human genetic maps. These maps are made from DNA sampled from individuals in a set of reference families where pedigree information showing the relationships among individuals in the reference families has been determined. The pedigrees consist of multiple families and multiple generations. Centres for human genetic mapping use the same reference families and share data so that information collected in one centre can be combined with information collected in other centers to construct a single genetic map.

Genetic chromosome maps show the relative locations of genes on a chromosome as determined by the frequencies of recombination between linked genes. The genetic map distances in such maps, however, do not necessarily represent physical distances between genes on a chromosome. Several methods permit geneticists to identify the physical locations of genes on chromosomes observed microscopically.

Deletion Mapping

In some animal species, chromosomes have been well characterised microscopically. Certain stains used in microscopy reveal banding patterns on chromosomes. Using banding patterns, researchers can identify not only chromosomes but chromosomal regions. A deletion is a mutation in which a segment of a chromosome is missing. Small chromosomal deletions usually include large amounts of DNA that may contain several genes.

Many deletions are lethal, particularly if they are large and include several genes. However, some deletions are small enough that they are not lethal when heterozygous, because the homologous chromosome carries copies of the genes in the deleted region. Most deletions are lethal when homozygous. Many deletions can be identified microscopically in chromosomes as missing regions in the banding pattern.

Deletion mapping is the identification of genes located in the deleted region of a deletion heterozygote. An individual that is homozygous for a recessive point-mutation allele in the gene of interest is crossed with an individual that is heterozygous for a deletion. If the gene is within the deleted region, half of the F_1 progeny display the recessive phenotype. Appearance of the recessive phenotype in half of the progeny is a good indication that the gene lies within the deletion and can be assigned to that chromosomal region.

An example of this is the *white* locus in *Drosophila melanogaster*. Several deletions that are homozygous lethal in females and hemizygous lethal in males have been found in the *white* region on the X chromosome and can be readily identified in the polytene chromosomes as missing bands. When female flies that are heterozygous for a deletion are crossed with flies with a recessive allele at the *white* locus, half of the female progeny display the mutant phenotype.

Somatic Cell Hybridisation Mapping

Somatic cell hybridisation is the fusion of human cells with rodent cells growing in culture. A fused cell initially contains all the rodent and human chromosomes. However, as the cell prepares to divide, the rodent chromosomes replicate and enter mitosis more quickly than the human chromosomes. As the cells divide, some of the unreplicated human chromosomes are lost. After several cell divisions, a cell line can be grown from a single cell. Chromosomes are lost at random, so a cell line may contain a particular combination of human chromosomes, and another cell line may contain a different combination.

The genes on human and rodent chromosomes are expressed in the cell lines. In some cases, enzymes differ enough in rodent and human cells that they can be distinguished from each other. When a cell line lacks a particular human enzyme, the gene for that enzyme must be located on a chromosome that is missing from the cell fine. When a cell has a particular human enzyme activity, the gene for that enzyme must be located on a chromosome that is present in the cell line.

Deletions are more frequent in human chromosomes, in somatic hybrid cell culture. Seletions can be induced in hybrid cell lines by treating them with ionising radiation. When a human chromosome is known to carry a certain enzyme gene, and a portion of the chromosome is deleted, the enzyme activity is lost if the gene is located within the deletion.

If the enzyme is not lost after the deletion, then the gene that encodes the enzyme must he outside the deleted region. Geneticists have used somatic cell hybridisation and deletion mapping to determine the physical locations of over 2, 000 genes on human chromosomes.

For many years, somatic cell hybridisation maps were the only detailed human genetic maps available, and they provided much information about the structure and organisation of human chromosomes.

DNA marker mapping and *in situ* hybridisation mapping have now moved to the forefront of human genetic mapping because researchers can map more genes in shorter periods of time with these methods than with somatic cell hybridisation.

Physical Chromosome Mapping

Genes can be mapped to their physical locations on the chromosome using deletion mapping or in situ hybridisation. Although the order of genes in genetic maps corresponds with the order in physical maps, genetic map distances do not necessarily correspond with physical map distances. A labelled probe that contains some of the DNA sequence from a gene can be hybridised with its complementary sequence in the DNA of a chromosome. The probe can then be detected at its position in the chromosome to reveal the physical location of the gene on the chromosome. *In situ* hybridisation to polytene chromosomes in certain insect species has been routine for many years. The development of fluorescent *in situ* hybridisation (FISH) has made it possible to identify the physical locations of genes in human chromosomes.

Physical chromosome maps correspond with genetic chromosome maps for the order of genes on the chromosome, but genetic map distances do not correspond closely with physical map distances. Varied levels of condensation within chromosomes observed microscopically and variation in crossover frequencies throughout a chromosome are two factors that contribute to the lack of correspondence between physical and genetic map distances. Genetic map distances may also vary significantly among species. In yeast, each cM unit averages about 3, 000 nucleotide pairs, whereas in humans each cM unit averages about 1,000,000 nucleotide pairs. Therefore, we cannot assume that genetic map units correspond to a constant physical distance.

PRACTICAL APPLICATIONS OF GENETIC MAPS

Genetic maps that include DNA markers are used routinely in medicine and agriculture. DNA markers that are coupled to an allele of an important gene can be used to identify the allele and track its inheritance. DNA markers can also be used as a starting point to find and clone the DNA of a gene that is closely linked to the marker.

Tracking Allele Inheritance with DNA Markers

The DNA sequence of the gene can be used to design a molecular marker that can detect the mutant allele directly. For example, the *HBB*S* mutant allele that causes sickle-cell anemia can be detected directly with RFLPs. However, until at least a portion of the gene has been cloned, it is not always possible to detect the gene directly using DNA markers. In these cases, a polymorphic DNA marker that is closely coupled to a mutant allele can be used to track the inheritance of the allele.

The mutant allele responsible for Huntington disease is a good example. In 1983, before the gene responsible for Huntington disease had been cloned and sequenced, an RFLP marker was identified that was closely coupled to the mutant allele. Clinical geneticists used this marker to determine whether people who had a parent with Huntington disease were heterozygous carriers of the allele. The gene responsible for Huntington disease has been cloned and sequenced, so a DNA test that detects mutant alleles directly, instead of a coupled marker, is now available.

The genes for most major genetic disorders have been cloned and sequenced, and DNA tests that detect mutant alleles directly are now available. However, until a gene has been cloned and sequenced, a coupled DNA marker is usually the best choice for tracking the inheritance of alleles of that gene. Plant and animal breeders routinely track important alleles, with coupled DNA markers.

After a genetic map consisting of DNA markers has been established in an important agricultural species, plant and animal breeders can quickly map genes of economic importance and find DNA markers that are closely coupled to alleles of those genes. Once closely coupled DNA markers have been found, it is possible to select for the marker and, in so doing, indirectly select for the allele.

Many agricultural plants are susceptible to certain diseases. In most cases, the best method for controlling the disease is to use plants that are genetically resistant to it. When disease-resistant plants are used, there is no need to fight the disease with chemicals that may be expensive and potentially harmful to people or the environment. Plant breeders must screen plants for resistance to common diseases and eliminate all that are susceptible in early generations.

To determine whether or not a plant is susceptible to the disease, breeders usually cultivate the disease organism and inoculate the plants with it. Sometimes this must be done in the field, and field testing may pose unwanted risks because the disease-causing organism may be released into the environment, where it may infect other plants.

However, if polymorphic DNA markers that are closely coupled to the alleles that confer resistance can be identified, DNA can be extracted from a small amount of leaf tissue in plant seedlings, and susceptible and resistant plants identified by testing for the DNA marker. The plants do not need to be inoculated with the disease-causing organism. Large numbers of plants may be screened in this way, thereby avoiding the risks and expenses associated with large-scale inoculation. A DNA marker that is closely coupled to an allele can be used to identify nearly all individuals that carry the allele. A few individuals, however, may be misclassified by detection of the DNA marker because a crossover separated the DNA marker from the allele.

Misclassification can be prevented by the use of two flanking DNA markers, markers that are located on either side of the allele. When flanking markers are used, only double crossovers separate both markers from the allele. Often flanking markers are so close to the allele that interference prevents double crossovers, and the allele can be identified with complete accuracy. This form of selection, in which genetic markers are used to indirectly select for a coupled allele, is called marker-assisted selection.

Chromosome Walking and Jumping

It is better to have a DNA clone of the gene itself, when tracking the inheritance of genes with DNA markers. It is possible to use the linked DNA marker as a starting point to work along the DNA and eventually arrive at the gene. Indeed, some important genes have been cloned in this way using procedures called chromosome walking and chromosome jumping.

Chromosome walking is the reconstruction of a long DNA sequence from many cloned segments of DNA by identifying overlapping clones. One done, whose position on the chromosome is known, may hybridise with a second clone whose position is not known. The second clone has a portion of its sequence in common with the first clone, so their DNA sequences overlap. Part of the sequence in the second done may extend beyond the limits of the first done. Once the correct position of the second clone ILS is established relative to the first done, the second done can be used to identify a third clone with an overlapping sequence. By repeating the process clone by clone, it is possible to 'walk' along a chromosome from a closely linked DNA marker locus to the target gene locus, without knowing the gene's sequence.

Chromosome jumping makes it possible to cover large regions of DNA from a clone with a known position to a second clone whose position is unknown. It can help speed up the process of chromosome walking. Chromosome jumping can detect two linked segments of DNA that do not overlap and are separated by up to tens of thousands of nucleotide pairs. Chromosome jumping is especially valuable when the distances between a linked marker and the target gene are large.

For chromosome jumping, DNA is digested with a rare cutting restriction enzyme, an enzyme that cuts DNA at an eight-nucleotide target sequence. On average, an enzyme with an eight-nucleotide target sequence cuts every 65,536 nucleotide pairs.

The DNA fragments generated with a rare cutting restriction enzyme are usually tens of thousands of nucleotide pairs long. The large fragments are mixed with a gene for a suppressor tRNA, and the fragments are treated to encourage their arrangement into circles.

Some of the circular DNAs contain large DNA fragments connected to the suppressor tRNA gene. The mixture is treated with the restriction enzyme EcoRl, which does not have a cleavage site in the tRNA suppressor gene. Since the circular DNA fragments are so large, they usually have several EcoRl cleavage sites. The enzyme trims the DNA on either side of the tRNA gene, leaving the tRNA gene flanked on either side by the ends of the large fragment. This fragment is cloned into a phage vector with a nonsense mutation that is suppressed by the tRNA suppressor gene.

The tRNA suppressor gene acts as a selectable marker and allows only those phages that contain the tRNA suppressor gene to replicate. If one of the ends of the cloned DNA hybridises to a clone that has been positioned in the chromosome walk, then the DNA on the other side of the tRNA gene should hybridise to another clone that is linked to the first clone, but at a large distance. The DNA on the other end can then be used for continued walking or for additional jumps.

REFERENCES

Creighton, H.B., andMcClintock, ., 'A Correlation of Cytological and Genetical Crossing-over in Zea Mays', *Proceeding of National Academy of Sciences*, USA 17: 492-497, 1931.

Crow, J. E., 'Mapping Functions', *Genetics* 125: 669-671, 1931.

Lindesley D., and G. Zimm., *The Genome of Drosophila melanogaster*, San Diego Calif.: Academic Press, 1992.

Ruddle F. H., and Kucheralapati, E.S., 'Hybrid Cells and Human Genes', *Scientific American* (Jul): 36-44, 1974.

Sturtevant A. H., *A History of Genetics*. New York: Harper and Row, 1965.

Sturtevant A. H., 'The Linear Arrangement of Sic-linked Factors in Drosophila, as Shown by Their Mode of Association', *Journal of Experimental Zoology* 14: 43-59, 1913.

5

Human Genome Research

The US Human Genome Project (HGP) composed of the DOE and NIH Human Genome Programmes, is an effort to understand the genetic instructions that make up a human. The human genome is the DNA that resides within every human cell. By sequencing the genome, scientists will identify the location and composition of all of our 100,000 or so genes. The genes carry information for making the proteins that direct the make up of each human. Among other things, proteins control human development, physiology, and resistance to disease. Knowledge of gene sequences is tremendously beneficial to scientists studying all facets of living things. To facilitate the future interpretation of human gene function, parallel studies are being carried out on selected model organisms.

As we usher in the post-genomic era, we are faced with an explosion of new information, which is leading to dramatic changes in the way we are able to study and manipulate life. On June 26, 2000, President Clinton and the leaders of the HGP and Celera Genomics announced the completion of a rough draft of the human genome, meaning that scientists have nearly finished determining the DNA sequence for all 23 pairs of chromosomes of the human species. The completion of the human genome project, and the sequencing of other organisms' DNA, is widely regarded as a turning point in biology and medicine.

Humans have a relatively large and complex genome, and so obtaining the complete sequence information has taken a relatively long time. During the time that the human genome was being sequenced, the smaller genomes of dozens of viruses and bacteria, including many human pathogens, have been completely sequenced.

Genomic sequences have also been obtained, or are being obtained, for several organisms that are widely used in basic biological research. These organisms include baker's yeast (*Saccharomyces cerevisiae*), the roundworm (*Caenorhabditis elegans*), the fruitfly (*Drosophila melanogaster*) and a small plant related to mustard (*Arabidopsis thaliana*).

The structure of our DNA, and the code that translates proteins from genes is virtually identical across all life forms, and we all share many of the same genes. Since this similarity, the knowledge that comes from studying non-human organisms is applicable to humans.

Scientists can also compare the genome sequences between organisms to learn more about what makes us different. For example, comparing DNA sequence information between organisms tells us more about the evolution of life. These molecular data can be used with the physical data from fossil records to give us a more complete understanding of the history of life on earth.

Each human being has a unique genomic sequence, except identical twins. It is the differences in our genomic sequences that give rise to the diversity of the human species. Our genetic differences make us unique in our appearance, our skills and interests, and in our physiology and biochemistry. The better we know ourselves, the easier it will be for us to maintain our health. Some of our genetic differences affect the way we react to drugs. Some tremendously beneficial drugs are ineffective, or even dangerous, in some people.

Millions of people are hospitalised, and hundreds of thousands die every year due to adverse drug reactions. Knowledge about which genes affect a person's response to a drug, called pharmacogenomics, can eventually allow doctors to prescribe the right drug at the right dose, all the time. Some of our genetic differences put us at a high risk of developing cancer, or heart disease, or depression or alcoholism. The knowledge of these genetic variations can help to prevent, or treat, these diseases.

The effort to sequence the human genome was entirely funded by public money. However, because this information has potential applications in the medical and pharmaceutical industries, private companies also began to sequence large parts of the human genome. Note that public (Human Genome Project) and private (Celera Genomics) organisations jointly announced the completion of the draft of the human genome.

One of the most controversial issues arising from the genome-sequencing project has been the question of who owns the information. Some feel that the patenting of genes will interfere with medical research in two ways. First, gene patents may raise the costs of further research. Second, scientists may become more secretive about their research in order to protect their rights to a potentially lucrative patent. Traditionally, scientists publish their research findings in scientific journal articles.

Technology available shortly after the start of the HGP in 1990 could have been used to attain HGP objectives, but the cost and time required would have been unacceptable. Since this, a major emphasis of the project's early years was to optimise existing methods and develop new technologies to increase DNA mapping and sequencing efficiency by 10 to 20-fold.

The genome now is being sequenced with technologies and methods that have evolved over the past 10 years. In the course of completing the sequence, a working draft of the human sequence was produced in 2000. The plan includes goals for further improving sequencing technology; studying human genome sequence variation; developing technology for functional genomics; completing the Caenorhabditis elegans and *Drosophila melanogaster* sequences and starting the mouse genome; studying the ELSI of genome research; studying bioinformatics and computation; and training genome scientists.

As part of the International Human Genome Project, vital and very active genome researches are also being pursued by researchers and science funding agencies inside and outside the United States. The DOE Human Genome Programme supports research projects at universities, the DOE Joint Genome Institute, DOE-owned national laboratories, and other research organisations are examples. Information obtained as part of the HGP will dramatically change almost all biological and medical research and dwarf the catalogue of current genetic knowledge.

Developed countries have handed over the human genome to private ownership together with the most triumphant hyperboles to boot, notwithstanding that it was mapped and sequenced at great public expense. A multi-billion bioinformatics goldrush is on, as private companies scramble to mine the public database for genes to patent and to assemble their own proprietary databases which are sold at exorbitant fees to subscribers. Beneath the hype, bioinformatics is a desperate attempt to turn the exponentially increasing amount of information into knowledge.

The human genome programme has dominated the scientific scene for the past ten years, raising hopes and fears in equal measure. We are already witnessing the resurgence of genetic discrimination and eugenics that have blighted the history of much of the last century. Bioinformatics suffers from the reductionist fallacy that knowledge will automatically arise once information is exhaustively listed.

Craig Venter, Director of Celera, claimed his discoveries could definitively cure cancer, thus securing Celera's place in the private investment market, while Francis Collins stressed that the real work is starting, thereby justifying the next round of major public finance. French Research Minister, Roger-Gerard Schwartsenberg, hailed the event as the victory of those who wanted knowledge to remain free.

Human genome has been sequenced separately and independently with major public finance, from the United States and the European Community. The US Government alone had earmarked $3 billion for the initiative. Celera's genetic maps would eventually be available on the Internet, and the company will claim royalties from any commercial pharmaceutical application of its discoveries. In contrast, the gene sequences and gene maps produced by the public consortium have been deposited regularly within 24 hours of completion in GenBank, a public database set up in the early 1980s when DNA sequencing began, access to which is totally free.

Celera kept its own human genome data secret while benefiting from free access to the public database throughout the period that the company was busy sequencing, thereby significantly reducing the time and effort needed to complete the task. Celera is not the only company stealing from HGP's Genbank. Others such as Incyte has mined the public data to help build its catalogue of genes and patents. At least 500 gene patents have already been awarded, while another 7, 000 have been applied for. Human Genome Sciences has won more than 100 gene patents and filed for roughly another 7, 000.

The G8 countries in line with the private companies, are downplaying the free access to the public human genome database on grounds that raw genome sequence is useless. During the quarterly meeting of Research Ministers of the G8 in Bordeaux at the end of June, to which Mexico, Brazil, China and India were invited, all agreed that DNA sequences—the fundamental data—must not be patented, in recognition that they are discoveries and not inventions.

This seems like a definite improvement over the previous situation in the United States where over four million patents on human genome sequences have already been granted, the majority of which are on short fragments of DNA with no known function. The US Patent and Trademark Office (USPTO) had tightened up the criteria for gene patents by issuing two new directives under section 101 utility, and section 112 written description requirements last December. Under the new utility guidelines, the USPTO is looking for specific utility and substantial utility. So, DNA fragments or express sequence tags (EST) will require a written description of their specific utility in order to be patented.

An industrial application may amount to no more than speculation based on similarity to gene sequences in the existing database. A notorious case involves the CCR5 gene patent awarded to Human Genome Sciences in the US this February. The company isolated the gene using automated computers to sequence it and software to determine that it belonged to a class of cell membrane receptors that pick up chemical signals in the body. Scientists at the Aaron Diamond AIDS Research Center in New York discovered that the AIDS virus requires the receptor to enter cells. A drug that can block the receptor would thus be a new weapon against AIDS.

Many patents have been awarded in association with condition X, where X is anything from cancer to criminality. There are already 740 patented gene tests on the market, among them BRCA1 and BRCA2, genes linked to breast cancer in women. Years after the tests were launched, scientists still do not know to what degree those genes contribute to a woman's cancer risk.

The US public GenBank holds sequence data on more than seven billion units of DNA, while Celera Genomics claims to have 50 terabytes of data in store, equivalent to 80 000 compact discs. The raw sequence data consist of monotonous strings of four letters—A, T, C and G that make up the three billion or so bases in the human genome. It is impossible to access the data or to make any sense of the sequences without special software.

Some software are developed and made freely available in the public domain, but the databases of private companies are provided to paid-up subscribers only. Incyte launched an e-commerce genomics programme in March that allows researchers to order sequence data or physical copies of more than 100, 000 genes on-line. Subscribers to the company's genomics database include drug giants such as Pfizer, Bayer and Eli Lilly.

Bioinformatics is a fusion of information technology with biology that promises to turn the raw genomic base-sequence data into knowledge for making even more lucrative new drugs. Bioinformatics is already a $300 million industry expected to grow to $2 billion within five years. One of the most basic operations in bioinformatics is searching for similarity or homology between a new sequence and one in the database, which allows researchers to predict the type of protein encoded and its function, thus enabling the sequence to be patented.

By understanding protein structure, it is possible to conduct searches for specific inhibitors and activators before carrying out actual biochemical experiments in the laboratory. Only one per cent of proteins so far has had their structures determined. Some bioinformatics companies cater to large users, aiming their products and services at genomics, biotechnology and pharmaceutical companies by creating custom software and offering consulting services.

Lion Bioscience, in Heidelberg, Germany, has a $100-million contract with Bayer to build and manage a bioinformatics capability across all of Bayer's divisions. Other firms target small or academic users. Web businesses such as Oakland, California–based Double Twist, and e-Bioinformatics in Pleasanton, California, offer one-stop internet shopping. These on-line companies allow users to access various types of databases and use software to manipulate the data. Large pharmaceutical companies have established entire departments to integrate and service computer software and facilitate database access across departments.

Proteomics is a part of bioinformatics focusing on when and where genes are active and on the properties of the proteins the genes encode. It attempts to make sense of the complex relationships between gene and protein and between different proteins and has so far also attracted hundreds of millions in venture capital. Proteomics has spawned a number of technical innovations, among which is the Gene Chip, developed by Affymetrix in Santa Clara, California. It consists of glass microarrays coated with cDNAs (complementary DNA) to identify which mRNA species are made. One microarray allows researchers to identify more than 60, 000 different human mRNAs.

Sequencing of the human genome is undeniably a technical feat comparable perhaps to landing on the moon. And it is difficult not to be caught up in a frenzy of speculation on what can be achieved as genomics joins forces with the latest in information and nanotechnology. According to John Bell at Oxford, within the next decade, 'predictive' gene testing will be widely used both in healthy people and for diagnosis and management of patients. Francis Collins, Director of the National Human Genome Research Institute in the US, has stated that the benefits of human genome mapping would include a new understanding of genetic contributions to human disease and the development of rational strategies for minimising or preventing disease phenotypes altogether.

During an industry conference held in Boston, senior executives from several of the world's leading genomics concerns agreed that genomics, with its promise of being able to show who will be predisposed to what disease, would eventually give rise to universal healthcare in the United States. This could happen especially if the defects in our genomes make us all uninsurable.

We could soon be able to catch deadly diseases in their earliest stages, when many are still treatable and even curable. And genomics also holds the promise of being able to deliver a bold new generation of drugs that will work more effectively with our individual genetic quirks. The bad news is that everyone will learn they are a walking time bomb, in one way, shape or form. The 'genetic mantle' currently put onto all diseases may prove to be like the emperor's new clothes. Most diseases are complex, and correlations between genes and disease are therefore weak. Associations between a disease and a genetic marker can occur by chance and some have proved to be spurious.

Although many disease-related genes have been mapped to regions of specific chromosomes, no clear markers for asthma, hypertension, schizophrenia, bipolar disorder, and other disorders have been found despite intensive efforts. Searches for susceptibility genes in breast cancer, colon cancer, rare early-onset forms of type II diabetes, and Alzheimer's disease have been more successful, but in each case these account for less than three per cent of all cases. That is because the risk of disease depends not only on other genes but also on environmental factors. The problem of identifying susceptibility genes is compounded when different combinations of genes are implicated in a disease, for it means that finding enough patients to serve as research subjects in a study will be extremely difficult.

The human genome sequence will enable geneticists

— to cure cancer.

— understand more about diseases and thereby to design better drugs.

— design customised cures based on our individual genetic make-up.

— prescribe an individual's lifestyle based on genetic make-up.

— diagnose all the bad genes that cause diseases.

— identify all the good genes responsible for desirable qualities such as longevity, intelligence, being slim and beautiful, good at sports, and so on.

— replace bad genes in gene therapy, including germline gene therapy.

— create genetic enhancement by introducing good genes.

— create designer babies and superior human beings.

However, in reality, the only concrete offering from mapping the human genome are the hundreds of patented gene tests. The high costs of the tests have prevented them from being used in cases where it might benefit patients in providing diagnosis. At the same time, those healthy subjects who have tested positive are likely to suffer from genetic discrimination and risk losing employment and health insurance. The value of diagnosis for conditions for which there is no cure is highly questionable. The claim to identify putative 'bad' and 'good' genes is also fuelling the return of eugenics, which has blighted the history of much of the 20th century.

This is exacerbated by the dominant genetic determinist mindset that makes even the most pernicious applications of gene technology seem compelling. A prominent band of scientists and bioethicists are actively advocating human genetic engineering, not just in gene therapy for genetic disease, but in positively enhancing and improving the genetic make-up of children of parents who can pay for the privilege, and have no qualms regarding human reproductive cloning either. In many ways, this is the most subtle form of hype for business to prosper.

Growth in bioinformatics is an admission of the vast realms of ignorance that separate the 100, 000 genes in the human genome from the living human being. It is also an acknowledgement that the genetic determinist paradigm, which has done so much to promote the human genome project, has failed miserably. There is no simple, linear causal chain connecting a gene to a trait, good or bad. Behind the hype is a desperate attempt to turn the exponentially increasing amount of information into knowledge that can pay off the heavy investments already sunk into the project.

Proponents of human genetic engineering see the creation of a genetic underclass to be inevitable, as those who can afford to pay for genetic enhancement will become gene rich relative to those who cannot afford to pay. The fallacy of genetic determinism is widely recognised. Genuine genetic diseases that can be attributed to single genes constitute less than two per cent of all diseases. And more and more geneticists are coming around to the view that even those are subject to so many other genetic and environmental influences that there is simply no such thing as a single-gene condition. For the rest, the association between the condition and the specific genes or genetic markers reduces to tenuous predispositions.

Predipositions to cancer conceals the fact that important environmental factors are left out of consideration. These include the hundreds of acknowledged industrial carcinogens polluting our environment. It is wellknown that the incidence of cancer increases with industrialisation and with the use of pesticides. Women in non-industrialised Asian countries have a much lower incidence of breast cancer than the women living in the industrialised West. However, when Asian women emigrate to Europe and the United States, their incidence of cancer jumps to that of the white European women within a single generation. Similarly, when DDT and other pesticides were phased out in Israel, breast cancer mortality in pre-menopausal women dropped by 30 per cent. The overwhelming causes of ill-health are environmental and social.

Genetic determinist approach of the human genome programme is pernicious because it diverts attention and resources away from addressing the real causes of ill-health, while at the same time stigmatising the victims and fuelling eugenic tendencies in society. The health of nations will be infinitely better served by devoting resources to preventing environmental pollution and to phasing out agrochemicals, rather than by identifying all the genes that predispose people to ill-health.

The effect of each gene depends not only on external environmental factors, but on the genetic background of all other genes in the genome. Individuals differ on average by one base per thousand in their DNA. This amounts to three million bases over the entire genome. As each gene is at least a thousand bases in length, it means that every gene will most probably be different. Assuming that only two variants exist in each gene, the number of different genotypes is already 3, 100, 000. In fact, hundreds of variants are typically found for each gene.

Consequently, every individual is genetically unique, except for identical twins at the beginning of development, before different genetic mutations can accumulate in each of the pair. That is why it is generally impossible to give accurate prognosis of even single gene diseases unless the genetic background is homogenous, as in an inbred laboratory strain of mice. The population in Iceland is thought to approach genetic homogeneity, which is why the company deCode Genetics has acquired the genetic database of Iceland's 270, 000 inhabitants, linked anonymously to medical records. The hope is to enable all the genes linked to a variety of diseases to be identified.

Unfortunately, the results will be valid for the Icelandic population only, and will not be transferable to other populations. Thus, mutations in the gene giving rise to cystic fibrosis among Northern Europeans is associated with quite another condition among the Yemans; while conditions diagnosed as *bona fide* cystic fibrosis in the latter population is associated with mutations in another gene altogether. Some geneticists, indeed, are beginning to think that better data for linkage to diseases might be found in genetically heterogeneous populations, such as those in Manhattan and London, rather than in homogeneous populations, such as those in Iceland.

The net effects of a gene are determined over all environments as well as over all genetic backgrounds, in recognition that both environmental and genetic interactions have to be taken into account. So, the most reliable data are those obtained in large populations which are as heterogeneous as possible genetically as well as environmentally. But the predictive power of such genetic data is always limited to population averages.

Up to 95 per cent of the human genome may be junk DNA, so called because no one knows what its function is. The same is true of all genomes of higher organisms. The rough draft of the human gene map announced in June 2000, is only 85 per cent complete for the coding regions only. It is difficult to see any definite strategy within either bioinformatics or proteomics that can pay off, either in terms of basic understanding the human organism as a whole, or in terms of miracle cures and wonder drugs.

Proteins encoded by the 100, 000 genes interact with one another, with the genes themselves, and small molecular weight cofactors and messengers. Those interactions vary in different cells and tissues at different times, subject to feedback from the environment. Feedback from the environment can alter the genes themselves, and hence the entire cascades of interactions involved. All that is the reality of the fluid and adaptable genome, which the moguls of genomics and bioinformatics have yet to come to grips with. The prospect of understanding the human being by a detailed description of its molecular parts is essentially nil. This reductionist fallacy has been exposed in different forms.

There is no unique correspondence between the states of the molecules within the system and the macroscopic condition of the organism, say, whether it is ill or well, simply because there are infinitely more molecular states than macroscopic states. Hence, a detailed description of all of the molecules, even if it were possible, will not enable one to determine the condition of the system as a whole.

We need a computer large enough to represent the states of all the molecules and their interactions in each cell, and fast enough to give a description of how they change in real time as the entire organism goes about its business of living. Current computation is unable to handle the dynamics of one single protein folding, even given all the information on the amino-acid sequence and the final shape of the folded protein. It takes the computer four hours to find a solution that is at best 70 per cent accurate; whereas the protein itself folds to perfection within a fraction of a second. What we need is a quantum leap to a new paradigm for understanding the organism as a coherent whole. Without that, human genome research will remain a scientific and financial black hole that swallows up all public and private resources without any return.

REFERENCES

Brown, K., 'The Human Genome Business Today', *Scientific American*, July, 50-55, 2000.

Ezzell, C., 'Beyond the Human Genome', *Scientific American*, July, 64-69, 2000.

Hayes, R., 'The politics of genetically engineered humans', *Exploratory Initiative on the New Human Genetic Technologies,* San Francisco, CA rhayes@publicmediacenter.org, 2000.

Holtzman, N.A. and Marteau, T.M., 'Will Genetics Revolutionise Medicine?' *The New England Journal of Medicine*, July, 343, 2000.

Howard, K., 'The bioinformatics goldrush'. *Scientific American,* July, 59-69, 2000.

6

Towards Genomics

Behaviour is the product of our most complicated organ, the brain. Most valuable for behavioural genetics will be the sequencing of multiple human genomes and identification of the several million DNA base pairs that differ among us. These DNA variations are responsible for the ubiquitous genetic influences on individual differences in behavioural dimensions and disorders. Genetic variation makes a substantial contribution to phenotypic variation for all behavioural domains. The best-studied areas are psychopathology, personality, and cognitive abilities and disabilities, all of which have been assessed by recent model-fitting. There are two striking findings.

The first is that nearly all behaviours that have been studied show moderate to high heritability. Second, although environment plays a role, its contribution tends to be of the non-shared type, that is, environmental factors make people different from rather than similar to their relatives. Such quantitative approaches, however, can no longer be seen as ends in themselves, and the information and techniques generated by the human genome sequence will be valuable in locating and identifying genes involved in behaviour. Although some genes that have major effects on behaviour have been identified, progress so far has been slow, partly because genes with effects large enough to be readily detectable by linkage are probably rare. For example, a recent genome scan in schizophrenia in almost 200 families effectively excluded a gene having a large effect from most of the genome. A further issue is the definition of phenotypes. Modern methods allow reliable diagnosis of disorders such as schizophrenia but there are several competing diagnostic schemes, raising the question of which one is search.

A complementary approach, allelic association, can detect genes that account for as little as one per cent of the variance in a trait. Until recently this approach was hampered by the lack of sufficiently detailed marker maps. Now, with the human genome sequence available, association studies should be able to locate QTLs with the aid of a greatly improved map based on hundreds of thousands of single-nucleotide polymorphisms (SNPs) and new methods of very high—throughput genotyping. Another new benefit is that much of the hard work of positional cloning has been replaced by cloning by computer.

Functional genomics and proteomics, where the focus is on gene products, their structure and expression, can be generally viewed as bottom-up strategies. But there are other levels of analysis at which we can understand how genes work. A top-down approach highlights the behaviour of the whole organism. For example, we can ask how the effects of specific genes unfold in behavioural development and how they interact and correlate with experience. This top-down, behavioural genomic level of analysis will complement the current functional approaches in the human species. Furthermore, genome sequences of other organisms will be especially important because of the great similarity of genes and gene organisation between.

The human genome sequence will revolutionise psychology and psychiatry. The most important impact will be on understanding the neurobiological basis of individual differences and achieving a better grasp of the etiology of diseases. The use of genomics as a path to drug discovery holds considerable potential. It is also probable that DNA testing will be used to predict which patients will respond to different drugs or be susceptible to particular side effects. However, there are two built-in limitations to this DNA revolution. The first is that all behaviour involves gene-environment interplay. The second is the unsolved question of the distribution of effect sizes of QTLs.

The probabilistic influence of genes on behaviour means that some of the ethical specters raised by the advent of behavioural genomics probably have little substance. Geneticisation is likely to increase the stigma of mental disorders.

POLITICAL ISSUES IN GENOME ERA

Sequencing of human genome heralds a new age in medicine, with enormous benefits for the general public. It will allow scientists to identify all of the series contributing to a given disease state, leading to a more accurate diagnosis and precise classification of disease severity. In addition, healthy patients can know the diseases for which they are at risk, giving them the opportunity to make beneficial lifestyle changes or to take preventive medications to protect their health.

There are many ethical, legal, and social concerns exist, despite the potential benefits. The US Congress recognised this early in the development of the publicly funded human genome project and so set aside approximately per cent per cent of the budget starting in 1990, to fund the ELSI programme (Ethical, Legal, and Social Implications of Human Genetics Research).

Initially, the ELSI programme focused efforts on four areas: privacy and fair use of genetic information, clinical integration of genetic technologies, issues surrounding research ethics, and public and professional education. Later, these goals were expanded to include studies of the societal impact of knowing the complete human genome sequence, the interpretation of genetic variations among individuals, integration of genetic technologies into clinical and non-clinical settings and the implications of genetic technologies for religious, philosophical, ethical, and socioeconomic concerns.

The most difficult issue is determination of proper balance between privacy concerns and fair use of genetic information. The growing number and use of genetic tests has many worried about discrimination due to inappropriate access to, and use of private genetic information. Francis Collins, Director of the National Human Genome Research Institute (NHGRI), has written. It is estimated that all of us carry dozens of glitches in our DNA—so establishing principles of fair use of this information is important for all of us. Without adequate safeguards, the genetic revolution could mean one step forward for science and two steps backwards for civil rights.

In 1998 the US National Centre for Genome Resources (NCGR) surveyed 1, 000 American adults and found that the 86 per cent believed that employers should not have access to a patient's genetic information, and 63 indicated they probably or definitely would not undergo genetic testing if they knew that insurers or employers could discover the results. However, members of the business community report that employment discrimination based on genetic information is currently very rare.

This situation has become more prevalent, and even a perception of genetic discrimination can seriously impede future progress. There are more barriers to achieving that era of personalised and preventive medicine than the scientific ones that have now been overcome. A key barrier is the fear that is pervasive in our society that genetic information will be used to deny health insurance or a job.

In US, federal laws such as the Americans with Disabilities Act and the Rehabilitation Act provide some protections against genetic discrimination in the workplace, but the scope of that coverage has not been tested in the courts. Former President Clinton recently signed an executive order barring genetic discrimination against employees in federal executive departments and agencies. US federal law does provide some protection against discrimination in health insurance. The Health Insurance Portability and Accountability Act of 1996

(HIPAA) bars a group health plan, or an issuer of a group health plan from using genetic information as a basis for implementing rules of eligibility for the plan or for setting premiums.

At present, 37 US states have laws regarding genetic discrimination and health insurance; 24 states have laws regarding genetic discrimination and employment. Although this patchwork of state laws affords some protections, it also contains loopholes. For example, definitions vary from state to state. One state may protect only DNA and RNA; another may extend protection to family history data and other medical information that could offer genetic clues. Ethical ambiguities are extended to the research methods used to gather the data in the first place. In large community studies, obtaining informed consent from every community member is often impractical. Studying groups of people within relatively small gene pools may have an unintentional stigmatising effect. Policies protecting confidentiality in research are crucial both to guard individual privacy and to promote advancement of the science.

Genetic information are being catalogued and maintained in forms of pathology specimens, blood bank donations, newborn screening samples, and research collections. Many countries maintain forensic DNA banks for use in criminal courts as well as commercial DNA banks. There have been efforts to create national genetic databases. For example, in December 1999, Iceland's parliament passed a bill allowing Decode Genetics, a biotechnology company, to combine all Icelanders' genetic, medical and genealogical information into one database to be sold to researchers. Estonian scientists are trying to create a similar genetic database and also to address concerns regarding access. Their goal is to include the genetic information, as well as other health and lifestyle data.

Genetic testing in the reproductive sciences is one of the most challenging areas of policy development. Research advances in this area have been remarkable, but are fraught with controversy. Couples considering pregnancy now have many options for genetic screening. In fact, those undergoing *in vitro* fertilisation may now opt to have their embryos genetically screened before implantation. This can be helpful to couples whose offspring are known to be at risk for an inherited disease.

Increased understanding of the human genome may ultimately result in the eradication of common diseases, but in the meantime we need to be on guard against potential misuse of genetic information. This is an emerging technology, and we should proceed with caution. The science is expanding at a breathtaking pace, and the overwhelming amount of new information puts governments under increasing pressure to pass legislation. Eventually every country must decide what genetic information should be protected, who will have access to it, and how it may be used. Governments must ensure that the public realises practical gains from their investment in generic technology, because much of the research is made possible by taxpayer-supported federal enterprises in partnership with academic and industrial institutions. The public must understand the new technologies so that unfounded fears will not develop and slow progress.

INTERPRETATION OF GENOME SEQUENCES

The first phase of the Human Genome Project constitute reports of the DNA sequence of the human genome and progress in sequencing the mouse genome. Analysis of these DNA sequences will reveal the inventory of genes used for building these organisms, as well as many regulatory elements that compose the instruction manual for converting the genetic parts list into organismal form and function. Research attention is now beginning to shift from problems of gene structure and genome organisation to questions of protein function and interactions, developmental and physiological pathways, and systems biology.

Many fundamental aspects of biological functions are not directly evident in DNA sequences. It is not unusual to discover a gene sequence about which little functional information can be deduced. Sequence analysis leads to no prediction of function for as many as 30 per cent of the genes in the human genome, and the inferred functions of most of the remaining genes have yet to be proven.

Due to the striking sequence similarities between humans and mice, discoveries in one species lead to strong inferences in the other. Laboratory mice and related species can make important contributions to functional genomics and identification of new models of human disease. Many spontaneous mutants have contributed profoundly to biomedical research and understanding of disease etiology and pathogenesis. The ability to make crosses between genetically defined strains, to work with large sample sizes, to engineer mutations in specific genes, and to generate mice with induced mutations facilitates identification of genetic variants of biomedical interest.

By including known single-gene mutants in surveys of mutagenised mice, induced mutations that modulate the mutant phenotype can be identified, as was done with great success in the discovery of naturally occurring variants that suppress disease traits in *Apc* and *Cftr* mutant mice. These mouse models reveal new drug targets for adenomatous *Polyposis coli* and *Cystic fibrosis*, as well as provide ways to evaluate potential therapeutics, predict treatment effects, and prioritise treatments for clinical trials. Remarkably, despite more than 100 years of research in mouse genetics, fewer than 5, 000 out of an expected total of 30,000 genes have functions attributed to them through direct experimental studies.

Recent progress in mouse genetics and genomics has provided proof-of-principle for large-scale studies to produce comprehensive collections of mouse mutations. These mutant mice are a permanent resource for future biomedical research, raising the possibility of attributing functions to every gene in the genome. International efforts to functionally annotate the DNA sequence of the mouse genome can only be achieved with a combination of phenotype and gene-based approaches in small and large-scale public and private projects.

Genome-wide mutagenesis, gene-trap, insertional mutagenesis, and large-scale genome alterations, as well as many kinds of targeted mutagenesis, are the methods of choice. Mutagenesis with agents such as ethyl nitrosourea (ENU) can be carried out in a high-throughput fashion in phenotype-based surveys. Although ENU causes numerous simple molecular lesions in each mutagenised mouse, its random action as a chemical mutagen can be a disadvantage in trying to study genes that warrant immediate attention because of their perceived importance. Other genes, such as the last gene in the genome to be mutated are difficult to obtain with agents that act randomly because of the properties of statistical sampling. In these cases, more directed strategies such as gene targeting and mutagenesis of embryonic stem (ES) cells *in vitro* are needed.

Gene targeting has traditionally been used to obtain loss-of-function mutations, but recent developments raise the possibility of obtaining many different kinds of mutations. Systematic analysis of the effects of generalised or tissue-specific mis-expression of gene products can also provide useful insights into gene function. Public and private programmes are under way in many countries, with funding from governmental agencies, private foundations and companies. These programmes involve survey for genes and traits affecting diverse biological traits and have already generated more than 1, 500 ENU-induced mutants, 150 gene-trap insertional mutations in ES cell, and roughly 4, 000 engineered mutants.

Leaders of these programmes now believe it is time to call for an International Mouse Mutagenesis Consortium (IMMC) that will systematically and comprehensively assign functions to every gene in the genome and will identify every gene that affects traits of high biomedical interest. This proposal is based on a public-private partnership with research groups in academia and industry working together to develop and apply genetic, genomic, phenotypic, bioinformatic, and computational methods to achieve the IMMC goals.

The long-range goals of IMMC are:

i) Produce at least one heritable mutation, in either ES cells or mice, in every gene in the genome.
ii) Identify every gene that affects key traits of biomedical interest.
iii) Establish an infrastructure for preserving and distributing mutant cells and mice.
iv) Enhance the informatics and database support for these functional studies.

The 10-year goals of the IMMC are:

i) Establish the International Mouse Mutagenesis Consortium. The mission of IMMC is to encourage the exchange of information and resources among participating mutagenesis laboratories and centres, to facilitate the distribution of information and reagents to the wider community, and to demonstrate to funding agencies that their local contributions are being leveraged into the larger international effort.
ii) Conduct a full-genome survey for key traits. The goal is to dig deep into the genome to find as many genes as possible that affect traits such as blood pressure, sleep, respiration, fertility, energy metabolism, thermal regulation, apoptosis, neural tube development, social behaviour, and olfaction. The traits selected for these studies should involve every major developmental and physiological system.
iii) Develop more precise and efficient phenotyping methods. Success of the IMMC depends on improved phenotyping.

Although many of the proposed goals can be achieved with existing methods, improvements are needed to reduce the amount of sample material for testing, to enable continuous monitoring in conscious, unrestrained animals, and to provide high-resolution *in vivo* imaging.

iv) Establish standard operating procedures. Phenotyping centres will publish on their websites the detailed protocols for each assay. These centres also need to coordinate the assays so that similar testing procedures yield similar results in different laboratories. Such standardised protocols are also important for sharing with research laboratories the assay conditions that were used in the original discovery and characterisation of each mutant.

v) Assign at least one function to as many genes as possible. The goal is to explore the breadth of biological functions. The challenge is to identify assays that will probe the extraordinary diversity of biological functions. This concerted effort will require ingenuity in devising many new screens for novel traits. These assays, which should be based on local research strengths and interests, will be applied to mutants on many different backgrounds. The need for divers assays represents an opportunity for large and small research laboratories.

vi) Determine the chromosomal location of every mutant gene.

vii) Develop efficient mutation detection methods to identify mutated genes.

viii) Use sensitised surveys to find enhancers, suppressors, and interacting pro reins for selected single-gene mutations.

ix) Establish a network of phenotyping centres with expertise in particular traits and biological functions.

x) Develop a centralised mouse phenotype database.

xi) Improve methods for preserving mutant mice in order to generate a permanent resource.

xii) Establish a cost-effective infrastructure for preserving and distributing mutant mice.

To make certain that mutant gametes, embryos, and mice are readily available, a network of resource centres is needed. The costs for these resource centres are substantial. Investments made today in mouse mutagenesis will be a foundation for future biomedical research.

Development of these comprehensive resources should be guided by principles that assure ready access and the broadest possible distribution. The value of the resources is their fundamental utility to enable subsequent research discoveries.

One of the most important principles of the Human Genome Project has been to assure the democratisation of access to tools and reagents. The same standards that have been accepted by the community with respect to access to structural genomic reagents such as clones, libraries, genetic markers, maps, and sequences should be applied to emerging functional genomics resources. Activities of IMMC will be coordinated under the auspices of International Mammalian Genome Society.

Being an IMMC member commits investigators to the common philosophy of open sharing of information and resources. The costs include mutant production with chemical mutagenesis, gene-trapping, gene-targeting, phenotype surveys and evaluation, mutant gene mapping and cloning, phenotyping centres, informatics and databases, and preservation and resource centres.

These estimates are based on the community experience in the various mutagenesis programmes that are ongoing in various countries. Mutant costs vary considerably, sometimes as much as 20-fold, depending on the nature of the assays used and differences in local expenses.

The following resources and technologies will enable the IMMC to pursue its goals:

i) *Improved methods to map mutant genes*: Improvements in technologies for genotyping single nucleotide polymorphisms (SNPs) may yield the necessary cost, scale, and efficiency to map large numbers of mutant genes.

ii) *Improved methods to identify mutant genes*: Although the mouse genome sequence provides the information and reagents for efficient identification of mutated genes, finding them by means of sequence analysis, especially with analysis of genomic DNA, is expensive. Methods that directly identify mutated genes, through chip-based or molecular technologies, would have great utility.

iii) *Improved phenotyping technologies*: Progress with DNA chips, micro-engineered machine systems (MEMs), nanotechnology, and imaging technologies provide enormous opportunities to revolutionise *in vitro, ex vivo,* and *in vivo* phenotyping methods by simultaneously increasing throughput and efficiency while reducing costs and the amount of sample material.

iv) *Improved informatics*: A website has been established that will act as a clearing-house for information on mutagenesis activities worldwide, including for each group their biological focus; genotype or phenotype screens; database access; and how to obtain mice, sperm, embryos, tissues, or cells. The site also provides a mechanism for individual and institutions worldwide to join the consortium. There is a need to ensure that all mutants, however generated are recorded with unique identifier in a common database. The Mouse Genome Database (MGD) provides the best means for achieving this goal, as each allele is listed as a separate entity with its own accession number. RAMC members are committed to entering validated mutants into MGD as soon as they are available for community access.

Unlike DNA sequence and genetic databases, there are no large-scale phenotype databases on which to model the databases for the mutagenesis centres, except perhaps those used to manage information for clinical and epidemiological studies. Several mutagenesis centres are exploring paradigms for these databases. In collaboration with the centre activities, MGD will serve as the community resource for integrating and unifying phenotypic information with genetic information about the laboratory mouse. We also need tools to assist in the integration of the relevant databases that will allow queries of all aspects of the biology of laboratory mice from DNA sequence to phenotypes.

The key challenges must be resolved for the proposed goals to be achieved are:

i) *High-throughput DNA-based mutation detection systems*: Positional cloning has been a successful approach for identifying mutant genes, and the rate of success is expected to increase dramatically with the availability of finished genome sequences for humans and mice. However, if a laboratory could positionally clone one gene per week, they would only identify 52 genes after a year of hard work. Even if one gene could be cloned per day, an extraordinary rate with existing technologies, only 365 mutants would be cloned per year. With perhaps 30,000 genes in the mouse genome, a network of 100 cloning centres would be required. Considerable improvements are obviously needed to identify the mutated genes more efficiently.

ii) *Functionally characterising trapped and engineered mutations*: In gene-based strategies, mutations can be obtained in specific genes of interest. The challenge with these gene-based methods is that the phenotypic consequences of the engineered mutations can be difficult to predict or detect. The natural world requires individuals to respond, from birth until death, to disease, infection, weather, social interactions, and many other factors, conditions, and processes that are not encountered in the simplified world of a laboratory mouse colony. These responses require coordinated action of diverse developmental, neurological, physiological, metabolic, and immunological systems. Existing technologies and phenotypic assessment paradigms are often inadequate to assess the subtleties of the biological responses to these challenges.

There are many other important genetic research activities deserve increased research support. These include the genetic and phenotypic analysis of naturally occurring polygenic traits, which is occurring within the Phenome Project, as well as studies of quantitative traits (QTLs) with chromosome substitution strains, deletion strains, and balancer chromosomes. Variation in naturally occurring traits is an important area, in part because many are models of human diseases. To date, however, finding QTLs has been difficult and additional research support is urgently need to map, clone, and characterise these genes in a more facile, high-throughput manner.

For the research biologist, it is easy to think about the advantages of having the sequence of every gene of potential interest, but another thing altogether to think about how to find all of them and to validate their identities and structures. The use of genome sequences to solve biological problems has even been afforded its own label for better or worse, it is called functional genomics. This new way of doing biology means some real changes, many of which are well under way in the community.

HGP has become familiar with the use of the full genome sequence in investigations of gene expression patterns and controls, protein-protein interaction networks, and other biological problems, since the publication of the *Saccharomyces cerevisiae* genome in 1996. These investigations are marked by a global point of view that was simply not possible before HGP had the sequence. Although HGP still do not know the function of about a third of the yeast genes, it do know that all possible protein and RNA participants in cellular function are encoded in the sequence it has.

From a picture of the software of the single cell, we can look to the future when researchers will begin building, with a fine a degree of resolution, an integrated view of the universe of cell-cell interactions, differentiation, and development from single cell to organism. The availability of complete sequences ɗ *Drosophila melanogaster*, *Caenorhabditis elegans*, and *Arabidopsis thaliana* is already beginning to revolutionise such studies, and this list may soon include significant sequence from other biological models of metasoan development.

The average number of alternates spliced from the transcript of a single mammalian gene might be in the range of two to three or more. As the present sequence yields estimates of about 30,000 genes, this would give an estimated 90,000 or more distinct proteins encoded by the human genome, without considering proteolytic processing or post-translational modifications. Thus, the complexity of the mammalian genome relative to that of yeast still presents formidable technical obstacles. So how can the working biologist take advantage of all this new information and bring about the advances predicted?

The first step is to understand that the present form of the available sequence information of the human genome is not a complete, fully annotated inventory of the human genes in each chromosome. Nor is the available sequence a single continuous and exact sequence for each chromosome. The reported genome sequence is represented by a set of sequences that cover the genome in a statistical sense but have a very large number of interruptions and gaps.

Although the completeness and continuity will continue to improve, there are significant uncertainties when inferences are made from these data. The concept of the 'contig' is essential to our understanding of this limitation. A contig is a contiguous piece of sequence information inferred by assembling sequence reads from single reactions. The number of contigs reported in the sequence data and their spectrum of sizes are important parameters in the analysis of genes.

As of 12 December 2000, the public database at the US National Centre for Biotechnology Information (NCBI) reported that the largest contig in the entire available sequence was 28.5 megabase pairs (Mb) in size; there were 43 contigs larger than one Mb, 566 contigs between 250 kb and one Mb, and 1, 628 contigs between 100 and 250 kb in size. This represented a total of approximately 600 Mb in contigs larger than 100 kb. Half of the sequence lies in contigs 22 kb or smaller, though they can be joined to form larger contigs.

We must distinguish here initial sequence contigs, derived from sequenced clones, and merged sequence contigs, derived by merging sequence contigs from overlapping sequenced clones. Since Venter et al., assemble sequence contigs, not from sequenced clones, but from the entire collection of sequence reads, this distinction is not necessary in their report. Since the average gene is of the same order of magnitude or larger than many of the contigs, this means that a significant fraction of human genes are unlikely to be represented on a single sequence contig in these data sets. The likelihood of finding one of the largest genes, such as Titin (-250 kb in size with >200 exons) on a single contig is much smaller than for small, simple genes like the olfactory receptor genes, which average less than two kb.

In the near future, many genes will have to be synthesised from an inferred organisation of the contigs into a gapped mosaic of assemblies called scaffolds. This means that an even more important factor than continuity for using the sequence to construct models of genes is the uncertainty associated with positioning the contigs relative to each other. Ambiguities in order and orientation of the contigs will sharply increase the number of possible ways that the sequence can be fit together and will thereby obscure the actual gene structure.

Venter *et al.*, report that they built scaffolds by using the paired-end sequences of their plasmid clones to link together and orient sequence contigs. They could put together these chains of sequence contigs, in the right order and orientation and at known distances apart, because they used several, known sizes of plasmid clones for sequencing and always generated sequence pairs at known distances from each other.

The advantages of relying on these kinds of sequence data were substantial in the assembly process. One of these advantages is that sequence contigs could be linked in the proper orientation and distance from each other even when they could not be merged into a single contig.

Thus, the self-consistent assembly from these data would appear to have ensured a high level of order and orientation of contigs at every scale of length. It may not be possible to fully assemble genes that fall into these scaffold segments if a gene segment falls into an unsequenced gap, but the picture of the gene that emerges should be fairly reliable. A gene would look something like the picture on a reconstructed Grecian urn, with blank clay segments holding the places for the real, picture-completing fragments.

A critical parameter for gene assembly and analysis for the Venter *et al.* approach is the size and coverage distribution of scaffolds. The average scaffold length reported was more than a megabase, with 25 per cent of the genome in scaffolds of at least 10 Mb in size. As the average gap length between scaffolds was only two kb, this data set seems to represent a high level of coverage for gene analysers, with a high level of consistent order and orientation. IHGSC had built their scaffolds by linking sequenced bacterial artificial chromosomes (BACs). This will still leave some sequence contigs within the BACs of the scaffold unordered or unoriented. The Grecian urn analogy does not fit here because the sizes and shapes of the gaps are not well known and, in some cases, the pieces may be in backwards or in the wrong order.

The critical factor for the gene-analysing biologist is the degree of ordering and orientation of contigs within the BACs that were linked to make the scaffolds, which is difficult to estimate from the report. Relevant measures include a reported overall estimate in the range of 10 to 15 per cent for misordered or misoriented sequence contigs. The reported N50 length for sequence contigs was 82 kb, and the scaffold N50 was 270 kb. Direct comparison of these statistics with the averages from Venter *et al.*, are not meaningful. To understand some of the subtleties, the interested reader will have to venture further into the data on distribution of lengths and other important complexities, keeping in mind the differences between the processes of assembly. The effectiveness of finding genes by similarity to a given sequence segment is determined by a much simpler statistic, the total coverage of the genome by the collective set of sequence contigs.

As the overall coverage of the genome is virtually complete. There is a strong likelihood that every gene is represented, at least in part in the data. Thus, finding any gene by sequence similarity searches using sufficient sequence to ensure significance is almost always possible using the data published this week. Caution must be exercised, however, as the identification of the gene may still be ambiguous. This is because a highly similar sequence from a receptor gene from *Drosophila*, for example, could be found in several different, homologous genes, which may have similar or entirely different functions or are non-functioning pseudogenes.

In other words, common domains or motifs can be present in many different genes. The use of the approximate similarity search tool BLAST is probably still the best way to find similar sequences. The excellent primer at the NCBI site should be used to understand the nature of the growing armament of BLAST-based tools, as well as the sometimes subtle issue of statistical significance and the limitations of this kind of approximate algorithm.

For most purposes, the approximation used by the BLAST algorithm is irrelevant, but the user should be aware of the specific kinds of similarities that may be missed by each available form of the algorithm. For example, since certain kinds of interrupted similarities are ignored, the more widely separated two similar sequences are, the less reliable will be the assessments of statistical significance. Newer methods attempting to use the structural cues inherent in the coding sequence to detect similarities are pushing back the detection limits for significant similarity.

Enormous progress has been made in automating the identification of genes in genomic sequence. But, building accurate models of genes from the sequence still requires a lot of human, hands-on effort. The best models are built of genes whose full-length mRNA sequences are available. The RNA sequence can be used to thread together the exon structure of the gene from genomic sequence no matter where the pieces may reside—continuity, order, and orientation of the fragments are not essential to this process.

Of course, the presence of pseudogenes and highly similar repeats can defeat even this strategy. Nonetheless, this represents a strong argument for gathering much more full-length cDNA sequence data. There are two general approaches to gene finding. The homology-based methods include the use of known mRNA sequences as well as gene families and inter-specific sequence comparisons. The *ab initio* methods include detection of exons and other sequence signals, like splice sites, by various computational methods within the sequence being analysed.

The location and structure of the sequences involved in regulation and control stands as one of the most difficult annotation problems, in every gene model. Finding and dissecting these important sequence regions can be done in some cases by means of motifs known to be conserved in transcription factor-binding regions.

Inter-specific genome comparisons are one of the ways of getting at these regions, under the assumption that the regions will stand out as being conserved. New experimental methods like an array-based technique to locate genome-wide sites of action of transcription factors, will also make significant contributions to sorting out the cis-regulatory signals in the genome. A number of tools are currently available for automated annotation, but a discussion of their advantages and limitations in specific circumstances is beyond the scope of this viewpoint. Approaches that use a combination of statistical methods to recognise genes and gene features are prevalent. They are most effective, however, in finding genes, rather than modelling them accurately, and are usually used in concert with homology-based methods.

Factors that can have strong effects on the effectiveness of such algorithms include errors in sequencing and statistical biases like base composition. Noise in the data can sharply degrade performance, so draft sequence, in which the error rate is higher, can be markedly inferior to finished sequence for *ab initio* prediction. GENSCAN is a widely used piece of software for gene finding and prediction, but newer developments like Genie also look promising. Genie is a hidden-Markov model system that allows for the integration of information from different sources such as signal sensors; sensors of introns and exons; and alignments of mRNA expressed sequence tag (EST), and peptide sequences.

Other software tools, like gene-builder, glimmerm, f-genes, grail, and others, have also been reviewed recently.

Using the *Drosophila* genome as a primary example, the Genome Annotation Assessment Project provides a very useful analysis of progress and problems in eukaryotic genome annotation. A similar comparison has been done using the *Arabidopsis* genome. The two genome papers have used systems consisting of multiple tools to create their initial gene inventories. IHGSC used a system called 'Ensembl' that follows *ab initio* predictions by GENSCAN with mRNA, EST, and protein motif information comparisons for the initial predictions. It then uses a programme called GeneWise, which has been used on the *Drosophila* genome, to extend protein matches. In contrast, Venter *et al.*, report the development of a rule-based expert system for annotation they call 'Otto' that attempts to embed some human curatorial functions in software.

These annotation efforts in the community are being linked to new ways of visualising genomes with their annotation. Genome browsers that enable the reader to navigate through many levels of genome information are now available that take the first steps in this direction. Commercial firms are also beginning to market similar kinds of software and are likely to continue to develop sophisticated, user-friendly packages for these purposes.

In the future, when the annotation of the genome is complete, the information from the sequence will be indicated in agreed-upon terms that can be searched directly by the text of the annotation. Clearly, a combination of sequence similarity searching tools, *ab initio* methods, and annotation-based searches must be used by researchers for the foreseeable future.

The next stage of annotation will also require the integration of independent experimental information into the gene annotation. To fully explore the properties of complex, highly interactive systems, databases will need to have pointers that link a gene to other genes by a variety of causal interactions, such as gene product X has a binding partner Y, exerts control on the expression of gene Y through a cis-regulatory site, produces a metabolic product that interacts with the product of gene Y, or participates in the same or linked signalling pathway with the product of gene Y. The way to this future has been opened by the availability of sequence information.

GENE NUMBER

In a monumental work, Venter *et al.*, confirmed that there might be fewer than 30,000 protein-coding genes in the human genome. That a mere one-third increase in gene numbers could be enough to progress from a rather unsophisticated nematode to humans and other mammals is certainly quite provocative and will undoubtedly trigger scientific, philosophical, ethical, and religious questions throughout the beginning of this new century. By the same token, humans appear only five times as complex as a bacterium like *Pseudomonas aeruginosa*.

Although a significant uncertainty is still attached to this low number, it was not totally unexpected, after the downward trend initiated by the analysis of the first two complete human chromosomes, as well as two independent statistical studies, and the unexpectedly low Drosophila gene number. After the older C value paradox, we now have an apparent N value paradox on our hands. Neither the cellular DNA content in mass nor its gene content appears directly related to our intuitive perception of organismal complexity.

A human is more complex than a nematode from a mere 100 to near infinity. Those widely different opinions were mostly the result of the lack of an objective measurement of what HGP mean by biological complexity. Some only considered the diversity of cell types, others considered brain circuitry, and others went as far as including the cultural achievements of the human species as a whole.

Furthermore, any personal estimate of biological complexity K can be fitted to the gene number N, by arbitrarily choosing a suitable functional relationship $K = f(N)$: proportional: $K \sim N$, polynomial: $K \sim N^a$, exponential: $K \sim a^N$, or even factorial: $K \sim N$.

According to the simplest model, in which each gene is either ON or OFF, a genome with N genes can encode 2^N states. According to this model, the human species appears more complex than the nematode species. $2^{30,000}/2^{20,000} = 2^{10,000} = 10^{3000}$.

This very big number can indeed accommodate the most idealistic opinions about the uniqueness of human beings and their superiority over worms.

Since genes are not independently expressed but are redundant and/or co-regulated in subsets, and also because many of these theoretical transcriptome states would be lethal, the exponents in the above formula would have to be reduced by one or two orders of magnitude. However, gene expression exhibits more than two states. A trivial mathematical model can thus illustrate how a relatively small number of genes could be sufficient to generate a tremendous biological complexity.

Biological sophistication evolves through the development of more individually and finely regulated gene expression mechanisms, rather than a sheer increase in the number of genes. Accordingly, metazoan promoters do obey more intricate triggering rules than their microbial counterparts, by making a combinatorial use of an expanded repertoire of transcription factors. The vertebrate immune system is another example of a biological system capable of generating a quasi-infinite repertoire of specific responses, by using a simple combinatorial logic involving a few hundred different genes that are regulated in a relatively straightforward manner.

If 30,000 genes might be compatible with the perception of right human complexity, this number has still to be reconciled with the much higher number of mRNA species as inferred from various assemblies of expressed sequence tags (ESTs). Alternative polyadenylation is an obvious explanation for this discrepancy. However, the latest estimate only predicts about 39,000 different endings from 30,000 genes. Alternative splicing is the next mechanism that can be invoked and could account for up to 48,000 different cDNAs.

Combining the detailed probabilities of both mechanisms in a simultaneous and independent manner could account for a maximum of 66,000 total different transcripts. Using another approach, HGP mapped each of the 82,000 Unigene clusters to the available human genome draft sequence in GenBank and searched for any significant protein homology within a 20-kb interval around each recognised genomic location. This computer experiment left with more than 46,000 unigene EST clusters for which there was no evidence of protein-coding potential.

Aside from the artifactual contamination by intron sequences, the large excess of cDNA/EST clusters over identified protein-coding genes could thus be explained by two main factors: the presence of numerous alternative forms of protein-coding transcripts, together with a significant number of transcripts from uncharacterised regulatory genes not encoding proteins. Genes of the latter category are not detected by the current *ab initio* gene-finding programmes and are usually discovered by chance. The thorough investigation of the nagging discrepancy between protein-coding gene and apparent mRNA numbers might thus still reveal some important biological discoveries.

A small number of human protein-coding genes means that the potential of functional genomics may be realised more easily and faster than anticipated. In the last few years, an increasing number of researchers have been saying that the old and classical reductionist approach would be totally inadequate to figure out the function of all genes. Instead, they propose that complex genetic networks should be studied as a whole, using new theoretical approaches, according to the premise that nondeterministic and/or chaotic phenomena might govern, the functioning of the human genome.

Unfortunately, these new approaches are still poorly developed, and have no track record of significant discovery in molecular biology. In fact, with only 30,000 genes, each directly interacting with four or five others on average, the human genome is not significantly more complex than a modem jet airplane. Yet, it is rarely suggested that airplane behaviour is mostly nondeterministic and requires a systemic understanding. Use of simple hierarchical regulatory models in conjunction with the spectacular development of high-throughput analyses will again be sufficient to rather quickly generate most of the significant results in functional genomics.

About 10 per cent of human genes might correspond to potential drug targets related to diseases of socio-economical importance. With only 3000 candidate genes to work from, i.e., 30 for each of the top 100 companies throughout the world, the pharmaceutical industry is now facing a new challenge. If the high-throughput approaches are used, developing leads for all of these candidates should only take a few years of fierce competition.

The development of personalised treatments based on genomic polymorphisms and individual transcriptome patterns might thus quickly become a necessary driving force of pharmaceutical innovation by reporting the generation and mapping of 2.3 million new single nucleotide polymorphisms (SNPs).

COMPUTATIONAL BIOLOGY

Genomics researches are heavily rooted in engineering technology, from the capillary electrophoresis units used in large-scale DNA sequencing projects, to the photolithography and robotics technology used in chip manufacture, to the confocal imaging systems used to read those chips, to the beam and detector technology driving high-throughput mass spectroscopy. Further advances in materials science and nanotechnology promise to improve the sensitivity and cost of these technologies greatly in the near future.

Genomic research makes it possible to look at biological phenomena on a scale not previously possible: all genes in a genome, all transcripts in a cell, all metabolic processes in a tissue. One feature that all of these approaches share is the production of massive quantities of data. GenBank, for example, now accommodates $>10^{11}$ nucleotides of nucleic acid sequence data and continues to more than double in size every year. New technologies for assaying gene expression patterns, protein structure, protein-protein interactions, etc., will provide even more data. How to handle these data, make sense of them, and render them accessible to biologists working on a wide variety of problems is the challenge facing bioinformatics—an emerging field that seeks to integrate computer science with applications derived from molecular biology.

Bioinformatics is an inherently integrative discipline, requiring access to data from a wide range of sources. Without the underlying data, and the ability to combine these data in new and interesting ways, the field of bioinformatics would be very much limited in scope. For example, the widespread utility of BLAST for the identification of gene similarity is attributable not only to the algorithm itself and its implementation, but also to the availability of databases such as GenBank, the European Molecular Biology Laboratory (EMBL), and DNA Data Bank of Japan (DDBJ), which pool genomic data from a various sources.

Computational biology is a fundamentally collaborative discipline, owing its very existence to the availability of rich and extensive data sets for analysis, integration, and manipulation. Data accessibility and usability are therefore critical, raising concerns about data release policies—what constitutes primary data, who owns this resource, when and how data should be released, and what restrictions may be placed on further use. One core aspect of research in computational biology focuses on database development: how to integrate and optimally query data from genomic DNA-sequence, spatial and temporal patterns of mRNA expression, protein structure, immunological reactivity, clinical outcomes, publication records, and other sources.

A second focus involves pattern recognition algorithms for such areas as nucleic acid or protein sequence assembly, sequence alignments for similarity comparisons or phylogeny reconstruction, motif recognition in linear sequences or higher order structure, and common patterns of gene expression. Both database integration and pattern recognition depend absolutely on accessing data from diverse sources, and being able to integrate and reproduce these data in new formats.

Two challenges have emerged that could potentially restrict the advancement of bioinformatics research:

i) questions related to the appropriate use of data released before publication and

ii) restrictions on the reposting of published data.

Use of Data before Publication

Recognising the value of early data release for a wide range of studies, the Human Genome Project adopted a policy of prepublication data release, and many genome projects now adhere to similar rules. Since bioinformatics depends absolutely on the ability to integrate data from a wide variety of sources, it is to be hoped that other projects that generate genomic-scale data will follow a similar policy, because immensely valuable results can emerge from large-scale comparative studies of genome structure, micro-array data, protein interactions, and so on.

The success of such altruistic data release policies, however, requires that those who generate primary sequence data receive appropriate recognition and are able to derive intellectual satisfaction from their work. Rowen *et al.* have recently proposed treating unpublished data available on the web as analogous to personal communication, thereby establishing some degree of intellectual property protection. The difficulty with this approach comes in determining what types of analysis should require permission from the submitters, and what types of analysis can reasonably be prohibited.

The identification of individual genes of interest for further experimental analysis must be acceptable-perhaps even without the need for formal permission—otherwise, early data release serves no purpose at all. Conversely, second-party publication of raw, unpublished, sequence data posted on the web must be viewed as violating ethical standards—analogous to the verbatim plagiarism of unpublished results from a meeting presentation.

Where to draw the line in intermediate cases will ultimately depend on the intellectual contributions provided by the manuscript in question, and whether such work might reasonably have been expected to emerge in due course from those who generated the original data.

Such considerations of value-added are not terribly different from those normally applied during manuscript review, but require special consideration by reviewers and editors of the anticipated contributions from the original submitter. Experience with the *Plasmodium falciparum* genome project suggests that disagreements over what kinds of data and analyses are permissible for publication are sometimes attributable to the failure of second parties to adequately consider the interests and involvement of those generating the primary data.

More often, however, disputes are attributable to a lack of understanding: either on the part of biologists, who do not fully appreciate the long lag that may reasonably be expected between the first appearance of shotgun sequencing results and final sequence closure and annotation, or on the part of those generating the primary data, who may not fully appreciate the intellectual contributions of biologists/bioinformaticians.

One hopes that as the gulf between those engaged in the application of genomic technologies bioinformatics research, and laboratory analysis is bridged by understanding, these problems will diminish in importance. Increased acceptance of web-based release as a form of publication, as well as increased understanding of the nature of big science projects in biology, will also reduce tensions.

Reposting Publishing Data

The challenge of restrictions on data access is of a more fundamental nature, involving not just when bioinformatic analysis is permissible, but what kinds of analyses can be carried out. Today's publication of a draft analysis of the human genome by Celera Genomics focuses a spotlight on this question, because the primary data themselves are being released only through a private company that places restrictions on the reposting and redistribution of their data. Other genome-scale projects, including a recent analysis of protein-protein interactions in *Helicobacter pylori*, have placed similar restrictions on the reposting of primary data. The data on which Celera's analyses are based will be available for examination. But the purpose of insisting that primary scientific data be released is not merely to ensure that the published conclusions are correct, but also to permit building on these results, to allow further scientific advancement.

Bioinformatics research is particularly dependent on unencumbered access to data, including the ability to reanalyse and repost results. A genome-wide analysis and reannotation of additional features identified in Celera's database could not be published or posted on the web without compromising the proprietary nature of the underlying data. Nor could this information be combined with the resources available from other databases such as the information from additional species necessary for cross-species comparisons, or data from microarray and proteomics resources that would permit queries based on a combination of genome sequence data, expression patterns, and structural information.

The present state of genomics research would never have been achieved without the freedom to use information from GenBank/EMBL/DDBJ. The potential for restricting downstream analysis offers the prospect of making a wealth of proprietary data generated by private companies accessible to the research community at large, but this potential comes at a very great cost.

Five years ago, the Bermuda Conventions established a standard for the release of genome sequence data that has served biologists very well; we should consider carefully what precedent to establish for the next five years, as considerations of data-release and data-use policy are likely to have far-reaching implications for all of biomedical research. The postgenomic era holds phenomenal promise for identifying the mechanistic bases of organismal development, metabolic processes, and disease, and we can confidently predict that bioinformatics research will have a dramatic impact on improving understanding of such diverse areas as the regulation of gene expression, protein structure determination, comparative evolution, and drug discovery.

The availability of virtually complete data sets also makes negative data informative: by mapping entire pathways, for example, it becomes interesting to ask not only what is present, but also what is absent. As the potential of genomics-scale studies becomes more fully appreciated, it is likely that genomics research will increasingly come to be viewed as indistinguishable from biology itself. But such research is only possible if data remain available not only for examination, but also to build upon.

ANALYSIS OF GENOME INFORMATION

Comparison of protein sequences encoded in genomes is a potentially powerful tool for identifying the components of functional systems and reconstructing their evolution. Such comparisons allow researchers to transfer information from well-studied model organisms to poorly characterised ones and to draw functional and evolutionary inferences from the presence, absence, and relative abundance of genes coding for different types of proteins in the compared genomes. Programmed cell death (apoptosis) is one of the central cellular processes in development, the stress response, aging, and disease in multicellular eukaryotes.

Comparative analysis of the components of the apoptotic machinery have shown that many of the protein domains that perform critical roles in this system were already present in the common ancestor of animals plants, and fungi. From the functions of the extant proteins containing these conserved domains, it can be extrapolated that they participated in ancestral signalling pathways, including those for pathogen and stress responses.

The evolution of the programmed cell death system from such signalling pathways was probably driven by general kin selection during the emergence of multicellularity. This chapter briefly discusses the results of a comparative analysis of the nearly complete protein sets of *Homo sapiens, Drosophila melanogaster,* and *Caenorhabditis elegans*. Evolutionary engineering of the apoptotic system followed the same pattern as seen in other signal transduction and regulatory systems, particularly in eukaryotes, namely, the formation of a wide variety of protein domain architectures from a relatively small set of ancient conserved domains.

Therefore, HGP applied a domain-centred approach to the comparative study of this system in animals. First the occurrences of the individual domains in apoptotic proteins were enumerated as accurately as possible by using a sensitive sequence analysis method based on the information contained in the multiple alignments of the corresponding protein sequences. Second, the domain architectures of the apoptotic proteins identified in humans, flies, and nematodes were systematically compared.

A panoply of proteins with functions in almost every basic cellular process have been directly or indirectly linked to apoptosis, which is not too surprising because programmed cell death is a complicated series of events involving various cellular subsystems. Examination of the number of occurrences of several domains that perform central functions in apoptosis shows a marked expansion in vertebrates relative to insects and nematodes. The growth in the number of these domains detectable in humans was noticed in all functional categories of proteins that contribute to programmed cell death, but was particularly striking among the extracellular components of the apoptotic system, the intracellular adaptor domains that transfer the signal from the receptors to the executors of apoptosis, the BCL2 family of apoptosis regulators, and the NACHT family of nucleoside triphosphatases.

Several secreted ligands, primarily members of the tumour necrosis factor (TNF) family, directly induce apoptosis, in vertebrates. A single, previously undetected member of the TNF family was identified in *Drosophila* which suggests that this ligand was already present before the divergence of the coelomates. The TNF family proteins function through specific receptors (TNFRs) that contain multiple repeats of an extracellular cysteine-rich domain and an intracellular Death Domain (DD). Predicted receptors with a single copy of the cysteine-rich domain are present in *Drosophila, C. elegans,* and plants, but none of them has the same architecture as the vertebrate TNFRs, and accordingly these proteins cannot be considered TNFR orthologs.

Transmission of external cell death stimuli to executors of apoptosis, such as caspases, is largely mediated by several specialised adaptor domains. The most prominent apoptotic adaptors are the CARD, DED, pyrin, and DD that have a common fold with six et helices and probably evolved from a common ancestor before the divergence of the extant animal lineages. A conspicuous expansion in the number of distinct proteins that contain these domains, particularly the CARD domain, is seen in humans.

The pyrin domain, which was predicted during the present protein sequence analysis, appears to be vertebrate-specific. The pyrin domain was identified in pyrin, Asc (a CARD-domain protein), the interferon-induced protein 16, the AIM2 protein, a caspase from zebrafish, and some uncharacterised proteins that also contain the NACHT NTPase domain. The BCL2 family proteins are conserved in all animals and have been implicated in alteration of mitochondrial permeability resulting in leakage of cytochrome c and the triggering of apoptosis.

The antiapoptotic members of this family interact with the Ced-4-like apoptotic adenosine triphosphatases (AP-ATPases) and inhibit their function in caspase activation, whereas the proapoptotic family members apparently interact antagonistically with antiapoptotic forms. Consistent with the experimental data, Ced-9 from *C. elegans* and the poriferan BCL2 homologue cluster with the antiapoptotic members of the family in phylogenetic analyses, whereas the fly BCL2s cluster with the proapoptotic versions.

Thus, the differentiation between the proapoptotic and antiapoptotic members of the BCL2 family might have been established in the coelomates, whereas the ancestral form apparently functioned only in the antiapoptotic capacity. The vertebrates show a proliferation of both these versions, with extreme sequence divergence observed in several proapoptotic members such as Bid and Mill.

The BH3-only proteins, which share only a region of limited sequence similarity with the BCL2 family proteins have recently attracted considerable attention. The BH3-only proteins interact with antiapoptotic members of the BCL2 family via an amphipathic helix formed by the BH3 motif and inactivate them.

The trend toward diversification in vertebrates seems to hold among the BH3-only proteins because several proteins of this group have been identified in mammals, as opposed to only one found thus far in *C. elegans* (EGL-1). However, there is no statistically significant similarity between diverse BH3 proteins, nor is the motif itself prominent enough to allow reliable sequence-based predictions in genome-wide searches. The NACHT NTPases that appear to be a sister group of the well-characterised AP-ATPases, such as Apafl and Ced4-, have been identified as participants in diverse regulatory interactions including activation of the transcription factor $NF_{K}B$ and apoptosis regulation (CARD4/NOD1 and NAIP), transcription regulation (CIITA), and telomerase function (TP1). All animals encode a TP1 ortholog that probably regulates the telomerase function and may not be involved in apoptosis.

Analysis of the human protein set shows a major, previously undetected expansion of the NACHT NTPase family; all newly detected members are closely related to CIITA, NAIP, and CARD4. rather than to TP1. In mice, a locus containing several highly conserved NAIP paralogs affects the survival of the pathogenic bacterium *Legionella* in macrophages.

The additional human NACHT NTPases might function in the regulation of immune response and related apoptotic processes. The increase in the number of proteins containing apoptosis-associated domains is accompanied by diversification of their domain architectures, including the emergence of a considerable number of lineage-specific architectures, particularly in vertebrates. Together with the numerical expansion, this amounts to a major increase in the complexity of the apoptotic system.

The diversification of domain architectures and increase in overall complexity are particularly remarkable in the case of the DD-fold adaptor domains that contribute to many domain combinations unique to the vertebrates, in addition to those formed by the vertebrate-specific offshoot of this fold, the pyrin domain. The NACHT NTPases also show diversification of the domain architectures in vertebrates through the addition of a variety of domains, such as BIR repeats, CARD, and pyrin, to an ancestral core that consists of a NACHT domain and leucine-rich repeats.

Several proteins with no detectable homologous outside the respective lineages have also been implicated in apoptosis. These include the proapoptotic protein SMAC (Diablo) that is specific to vertebrates and three small proteins with a similar hydrophobic NH_2-terminal peptide-Reaper, Grim, and Hid in *Drosophila.* These appear to have evolved largely from compositionally biased, non-globular, or predominantly

α-helical proteins through selection of specific peptides for interactions with other proteins.

The evolution of the cell death pathways in animals can now be reconstructed from the wealth of genomic information. Several of the key domains of this system were apparently present in the common ancestor of the eukaryotic crown group. These ancient homologous of apoptotic proteins include enzymes such as the caspases, the predicted A20-like protease, AP-ATPase, NACHT NTPase, and the previously undetected apoptotic guanosine triphosphatase (AP-GTPase); adaptors such as TIR, BIR, and MATH and nuclear factors such as E2F, Rb, and signal transducers and activators of transcription (STATs).

Apoptotic proteins like BIR-domain protein survivin, appear to have an ancient function related to mitosis and cell cycle regulation rather than to apoptosis; this protein probably has been recruited for its antiapoptotic function only in the coelomates. Others, including proteins containing the TIR, caspase, and AP-ATPase domains, possibly interacted to form one or more pathways related to apoptosis and involved in pathogen or stress response. Participation of the MATH domain, the RING finger and possibly A20-like proteases in the ubiquitin signalling system suggests that these proteins as well as the ubiquitin-containing protein kinase IKK have been recruited for their functions in apoptosis from the ubiquitin-based pathways.

Apoptotic system resembling the core of the extant one appears to have emerged concomitantly with the origin of the metazoans. This event apparently was marked by the rapid divergence of the caspase-paracaspase protease family from a metacaspase-like ancestor, followed by the divergence of classical caspases and paracaspases. Another key early event in animal evolution was the emergence of the six-α-helical adaptor domain, which was soon followed by its diversification, the earliest split probably being between the DD and CARD domains, which are the only two domains of this class that apparently are present in all animals.

The direct apoptotic function of these domains in early animals remains to be ascertained, but even if they originally played a different role, the presence of DD and the previously undetected ZU5 domain in the netrin receptors and ankyrins in all animals suggests that these domains were already used in cytoskeleton and receptor-mediated signalling. Other components of the apoptotic system and related molecules that were probably present in the common ancestor of all animals include the BCL-2 family proteins, certain adaptors such as TRAFs and Tollip, the A20-like protease, the AP-ATPase, and the IRAK and DAP protein kinases.

The conservation of all these apoptotic components in animals suggests that a relatively simple, but complete, molecular machinery for programmed cell death had evolved before the divergence of the major animal lineages. Relative to the number of orthologs between nematodes and arthropods to the exclusion of vertebrates, vertebrates and arthropods share more orthologs of the apoptotic system components to the exclusion of nematodes.

The group of apoptosis-related proteins specifically shared by vertebrates and arthropods includes the transcription factors NFAT and NFKB that apparently have evolved from ancestral immunoglobulin (Ig) domain-containing transcription factors, such as OLF-1 (SPT23) or Su(H), and the signalling cascade associated with NFKB. This cascade minimally consists of the Toll-like receptors, adaptors (MYD88 and FADD), and protein kinases including NIK (NFKB-inducing kinase) and two paralogs of IKK.

The presence of TNF but the apparent absence of a TNFR in the common ancestor of insects and vertebrates suggests differences in the upstream portion of this apoptotic pathway. Another group of apoptosis-associated proteins that are shared by vertebrates and arthropods to the exclusion of nematodes is the CAD family, whose members regulate post-mortem DNA degradation.The domains and domain architectures present in vertebrates and insects but not in nematodes are indeed shared derived characters of the coelomate clade. This is compatible with the traditional view of animal evolution but not with the currently popular ecdysozoa model, which argues for a clade of molting animals including arthropods and nematodes. However, the alternative explanation-coordinated loss, in the nematode lineage, of multiple genes coding for proteins involved in several apoptotic pathways-cannot be entirely ruled out.

The prevailing theme in the evolution of the apoptosis-associated domains in the vertebrate lineage is the growth of complexity that is detectable across the entire range of the apoptosis-associated proteins and domains. In some cases, such as the BCL2 family, this is achieved primarily through duplication with limited diversification; on other occasions, the emergence of new domains through a more radical modification of preexisting ones, and reorganisation of protein domain architectures may be equally important.

To a large extent, the innovations in apoptotic and related cytokine signalling in vertebrates could have been linked to the evolution of the vertebrate immune system, with its several new cell types that require highly specialised regulatory pathways. Perhaps the greatest mystery in the evolution of apoptosis is the presence of homologues of several components of the eukaryotic apoptotic machinery in bacteria. At least two bacterial lineages, *Actinomycetes* and *Cyanobacteria*, encode a considerable repertoire of apoptosis-associated domains, including AP-ATPases, metacaspase-like proteases, NACHT NTPases, and TIR domains.

Some of the bacterial AP-ATPases are involved in transcription regulation and signalling, whereas the functions of the rest of these proteins remain unclear. However, it is almost certain that they are functionally connected, given the fusions of the metacaspase like domain and the TIR domain with AP-ATPases. The presence of the apoptosis-associated domains in the crown-group eukaryotes and in specific divisions of developmentally complex bacteria contrasts with their complete absence in archaea and in other bacteria and suggests a history of concerted horizontal gene transfer.

The direction of this transfer, however, is uncertain, and although acquisition of the corresponding genes from bacteria by early eukaryotes seems more likely—because the bacterial lineages probably had been fully established by the time of the emergence of the crown-group eukaryotes—the opposite model of a relatively late dissemination from eukaryotes to the bacteria cannot be dismissed. The principal conclusion from the comparison of the apoptotic system components and their homologues encoded in the sequenced eukaryotic genomes is the major increase in complexity in vertebrates relative to insects and nematodes. This is manifest both in a numerical increase of apoptosis-related proteins and in domain accretion, which leads to increasingly elaborate domain architectures within orthologous protein sets.

Only the genome sequences allow for a reasonably accurate quantitative comparison of the complexity of functional systems, including the apoptotic machinery, in different organisms and for a reasonably confident reconstruction of the ancestral systems. Moreover, the expansion of certain protein and domain families, such as the NACHT NTPases and the pyrin domain in vertebrates, became apparent only from the analysis of the nearly complete sequence of the human genome.

Any statements that a particular protein or domain is lineage-specific—that is, missing in other lineages—rely both on the completeness of a representative genome sequences from each of the lineages and on the assumption that they accurately reflect the gene complement of the entire lineage. With the completion of several eukaryotic genomes, the study of the functional systems of these organisms, including apoptosis, is entering the postgenomic era. However, to understand the origin and evolution of apoptosis at a more satisfactory level, we need more genomes from diverse branches of life.

DNA Repair Genes

DNA repair enzymes continuously monitor chromosomes to correct damaged nucleotide residues generated by exposure to carcinogens and cytotoxic compounds. The damage is partly a consequence of environmental agents such as ultraviolet light from the sun, inhaled cigarette smoke, or incompletely defined dietary factors. However, a large proportion of DNA alterations are caused unavoidably by endogenous weak mutagens including water, reactive oxygen species, and metabolites that can act as alkylating agents. Very slow turnover of DNA consequently occurs even in cells that do not proliferate.

Genome instability caused by the great variety of DNA-damaging agents would be an overwhelming problem for cells and organisms if it were not for DNA repair. The functions required for the three distinct forms of excision repair are described separately. These are base excision repair (BER), nucleotide excision repair (NER), and mismatch repair (MMR).

The BER proteins excise and replace damaged DNA bases, mainly those arising from endogenous oxidative and hydrolytic decay of DNA. DNA glycosylases initiate this process by releasing the modified base. This is followed by cleavage of the sugar-phosphate chain, excision of the abasic residue, and local DNA synthesis and ligation.

Cell nuclei and mitochondria contain several related but nonidentical DNA glycosylases obtained through alternative splicing of transcripts. Three different nuclear DNA glycosylases counteract oxidative damage, and a fourth mainly excises alkylated purines. Remarkably, four of the eight identified DNA glycosylases can remove uracil from DNA. Each of them has a specialised function. UNG, which is homologous to the *Escherichia* coli Ung enzyme, is associated with DNA replication forks and corrects uracil misincorporated opposite adenine. SMUGI, which is unique to higher eukaryotes, probably removes the uracil that arises in DNA by deamination of cytosine.

MBD4 excises uracil and thymine specifically at deaminated CpG and 5-methyl-CpG sequences, and TDG removes ethenoC, a product of lipid peroxidation, and also slowly removes uracil and thymine at G-U and G-T base pairs. The existence of multiple proteins with similar activities is a recurring theme in human DNA repair. Another illustration of this is the set of at least four adenosine triphosphate (ATP)-dependent DNA ligases encoded by three genes, with LIG3-XRCC1 providing the main nick-joining function for BER.

Until recently, only one endonuclease for abaisc sites had been found encoded in the *E. coli* and the yeast *Sacchromyces cervisiae* and three genes were predicted in the genome of the plant *Arabidopsis thaliana*. A second human gene, APE2, has recently appeared. Apparently this encodes a minor activity, as deletion of the major gene APE1 causes early embryonic lethality in mice. Repair of the DNA replication-blocking lesion 3-methyladenine is another case where the human genome is frugal.

In other organisms, several DNA glycosylases, unrelated at the primary sequence level, can remove 3-meA. Among them are Tagl of *E. coli*, AlkA of *E. coli*, and MPG in higher eukaryotes. Only the MPG enzyme has been characterised so far in the human genome. This is in contrast to at least two *alkA* and six *tag1* homologues found in *Arabidopsis*. However, like the genomes of other multicellular animals, the current human genome draft contains no obvious *tag1* and *alk4* homologues.

The human MGMT removes methyl groups and other small alkyl groups from the O^6 position of guanine. There are two such proteins (Ada and Ogt) in *E. coli*, but no additional homologues have been detected in the human genome sequence. MGMT resembles the COOH-terminal half of Ada. The NH_2-terminal half of *E. coli* Ada can remove a methyl group from a DNA phosphate residue.

HGP found no homologues of this region of Ada, and it remains unclear whether such backbone methylations are repaired in human cells.

Photolyases that can monomerise lesions induced by UV light such as cyclobutane pyrimidine dimers and photoproducts. The human genome has two CRY genes with similarity to photolyase sequences. These encode blue light photoreceptors involved in setting of circadian rhythms but not in photoreactivation of DNA damage.

HGP has not detected additional homologues of DNA repair photolyases in the human genome, confirming previous reports that photolyase activity is present in many vertebrates including fish, reptiles, and marsupials, but not in placental mammals. NER mainly removes bulky adducts caused by environmental agents. In *E. coli*, the three polypeptides UvrA, UvrB, and UvrC can locate a lesion and incise on either side of it to remove a segment of nucleotides containing the damage. Eukaryotes, including yeast and human cells, do not have direct UvrABC homologues but use a more elaborate assembly of gene products to carry out NER. For example, *E. coli* UvrA can bind to sites of DNA damage, whereas at least four different human NER factors have this property.

The formation of an unwound pre-incision intermediate in human cells requires two DNA helicases, XPB and XPD, instead of the single UvrB in *E. coli*, and there are dedicated human nucleases (XPG and ERCC1-XPF) for each of the two incisions, instead of the single UvrC in bacteria. S. *cerevisiae* encodes two additional gene products, Rad7 and Rad16, which are important for NER. No convincing homologues to these can be identified in the human genome, although Rad16 is a difficult case because it is a member of the amply represented Swi/Snf family of DNA-stimulated adenosine triphosphatases.

Organisms such as the fission yeast *Schizosaccharomyces pombe* have a second system for excision of pyrimidine dimers, initiated by a UVDE nuclease. The human genome apparently lacks a homologue of this nuclease and has no such backup system, consistent with the fact that cells from NER-defective xeroderma pigmentosum patients totally lack the ability to remove pyrimidine dimers from DNA.

The transcribed strand of active human genes is repaired faster than the nontranscribed strand in a transcription-coupled repair process known to involve the products of CSA, CSB, and XAB2. MMR corrects occasional errors of DNA replication as well as heterologies formed during recombination. The bacterial *mutS* and *mutL* genes encode proteins responsible for identifying mismatches, and there are numerous homologues of these genes in the human genome, of greater variety than those found in yeast, *Drosophila melanogaster,* or *Caenorhabditis elegans.*

Some of these proteins are specialised for locating distinct types of mismatches in DNA, some are specialised for meiotic recombination, and some have functions yet to be determined. In *E. coli,* the newly synthesised DNA strand is identified with the aid of the MutH endonuclease, which has no human ortholog. Strand discrimination in human cells may be signalled instead by the orientation of components of the DNA replication complex such as PCNA or by other factors not yet identified.

DNA double-strand breaks may be rectified by either homologous or nonhomologous recombination pathways. Particularly notable in the human sequence is the presence of at least seven genes encoding proteins distantly related to the single Rad51 of S. *cerevisiae* and the single RecA of *E.coli*. The latter proteins function in strand pairing and exchange during recombination. By comparison, four members of the Rad5l family have been found in the *Drosophila* genome and four in *Arabidopsis*. Homologous recombination in human cells is likely to involve branch migration enzymes and resolvases that are functionally analogous to the bacterial RuvABC system. Single-strand interruptions in DNA can be rectified by enzymes from the BER pathway. Enzymes of the PARP family, as well as XRCC1, temporarily bind to single-strand interruptions in DNA and may act to recruit repair proteins.

There are currently at least 15 DNA polymerases in humans, exceeding the number found in any other organism.

For repair of nuclear DNA, the main form of BER uses Pol b, whereas Pol d or Pol e are the main enzymes employed for NER and MMR. Genetic and biochemical evidence has implicated many of the newly discovered polymerases in the DNA damage response, but others may have specialised roles such as sister chromatid cohesion. REV3L, the catalytic subunit of DNA polymerase 4, illustrates how DNA sequence homology searches can yield unexpected results.

The DNA polymerase domain at the COOH-terminus of the human protein resembles *S. cerevisiae* Rev3, but most of the first 2, 000 amino acids are not present in the yeast protein. A second human gene highly homologous to 1, 200 residues in this region is encoded on the X-chromosome. It is premature to classify this as a DNA repair gene, but study of it is expected to shed light on the function of REV3L.

Human genome sequence has already markedly influenced the field of DNA repair. Many of the genes listed were discovered as investigators searched the expanding database for sequence similarity to genes discovered in model organisms. This approach will no doubt continue and new human genes will be identified as additional repair functions are identified in other systems.

One source that is likely to be fruitful is the genome of *Deinococcus radiodurans*. This bacterium has an exceptionally high resistance to DNA-damaging agents, especially ionising radiation, in comparison to other microorganisms. Some of the currently uncharacterised genes in *D. radiodurans* are expected to contribute to DNA repair, and it remains to be seen if there will be homologues of such functions in the human genome.

Sequence database also makes it increasingly straight-forward to use mass spectrometry fingerprinting to identify new subunits of repair protein complexes. In this sensitive technique, isolated proteins are digested with an enzyme such as trypsin, and the exact molecular masses of the resulting fragments are measured. Comparison of these fragments with a computer-simulated tryptic digest of each human gene product can unambiguously identify the protein. In addition, new genes will be found as novel biochemical assays are developed for various aspects of repair. For example, human cells can repair cross-links between the two DNA strands.

Interstrand cross-links are generated by natural psoralen compounds and their chemotherapeutic derivatives, by other drugs used for cancer treatment such as nitrogen mustards, and to some extent by ionising and ultraviolet radiation. Repair of such cross-links involves the NER genes and the XRCC2 and XRCC3 recombination genes and is predicted to involve the DNA polymerase POLQ. In addition, the sensitivity of cells from individuals with Fanconi Anemia (FA) points to a role for the FANC group of genes in cross-link repair.

New genes may be identified, for instance, involved in the repair of damage caused by lipid peroxidation. Other uncharacterised forms of DNA damage caused by reactive metabolites and catabolites may be found. For example, the genome is dynamic, and single-stranded regions are temporarily exposed during DNA replication and gene transcription. Positions that are normally protected by base-pairing within the double helical structure are then vulnerable to group-specific reagents, creating new classes of lesions. Alkylating agents can form the cytotoxic lesions 1-methyladenine and 3-methylcytosine in single-stranded DNA, and new repair strategies may be needed to remove such lesions.

DNA metabolic processes need a close connection with proteins that allow chromatin remodeling or disassembly, because, DNA is assembled into several levels of ordered chromatin structure. Several human chromatin remodeling complexes are known, for instance, that allow and control access to DNA during gene transcription.

The great majority of enzymological DNA repair studies to date have worked with naked DNA, but chromatin presents a substantial barrier to recognition of DNA damage. It is expected that human protein complexes will be found that are dedicated to DNA repair and recombination, facilitating access of DNA repair enzymes to the genome.

Three-dimensional structures of DNA repair proteins are being determined at an ever-increasing pace. The functionally related SMUG1, TDG, and UNG enzymes show little or no primary sequence homology yet have common structural folds and belong to a single protein superfamily. As the structures of new protein folds are documented, more members of DNA repair enzyme families are likely to be found with the aid of three-dimensional structure prediction models.

Knockouts of DNA glycosylases in mice have unexpectedly mild consequences by comparison with budding yeast and *E. coli* models. This implies that more backup systems exist, probably because endogenous damage presents a more frequent problem for larger genomes.

As the genes from the human genome sequence continue to be catalogued, studying the activity of the protein products will become increasingly important. More effective methods for rapid expression of active proteins will be required to test for possible functions. An alternative approach is to selectively inactivate individual proteins *in vivo*.

An efficient method for selective proteolytic destruction has been successful in budding yeast and should be extendable to mammalian cells. Alternatively, systematic interference with gene expression with the use of inhibitory RNA molecules, as employed successfully in C. *elegans*, is proving to be a powerful way to dissect gene functions. Intense activity is being devoted to understanding how DNA damage transmits signals to the cell-cycle checkpoint machinery and to the monitoring systems that control cellular apoptosis. There is recent progress on this complex extended network, which involves damage recognition factors, protein kinases, and transcription factors such as p53.

Attempts are already being made to obtain an integrated picture of DNA repair with regard to signalling. The subject is of great interest as some inherited human syndromes associated with sensitivity to DNA-damaging agents result from loss of functions such as ATM, which is involved in damage sensing.

Clinical applications relating to human DNA repair genes are certain to emerge. Tumour cells often acquire resistance to therapeutic drugs or radiation. Genomics approaches such as array technology will be used to define any DNA repair genes that may be over-expressed in this context. Furthermore, it will be important to find ways to specifically inhibit DNA repair in these resistant cells by targeting the key enzymes.

Genetic polymorphisms in relevant repair genes will be identified and efforts made to correlate them with effects on activity of the respective proteins, with response to particular therapies and with clinical outcomes. Although a number of polymorphisms in DNA repair genes are being reported, there is presently little functional information on the consequences of the attendant amino acid changes. It will be important to find out which polymorphisms actually affect protein function and then concentrate on these in epidemiological and clinical studies. For example, homozygosity for a particular polymorphism in the DNA ligase subunit XRCC1 is associated with higher sister chromatid exchange frequencies in smokers, suggesting an association of this allele with a higher risk for tobacco- and age-related DNA damage.

With the use of gene and protein array techniques, it should be possible to compare expression profiles of DNA repair genes in normal and tumour cells-information that could eventually lead to individually tailored therapies with chemicals and radiation. For example, tumours with low levels of NER should be more susceptible to treatment with cisplatin. In experimental systems, MMR-deficient cells are highly tolerant to alkylating chemotherapeutic drugs. MMR-defective tumours such as those found in hereditary nonpolyposis colon cancer may be resistant to treatment with such agents.

Variation in DNA repair gene expression is epigenetic in origin and has been found for instance with MGMT and MSH6. The MGMT gene promoter is often methylated in gliomas, resulting in suppressed expression that can be associated with an improved response after tumour treatment with an alkylating agent. The complete human genome sequence now allows the definition of promoter regions so that the DNA methylation status of relevant CpG islands can be investigated readily.

DNA repair, especially repair of oxidative damage, has often been suggested as a relevant factor in counteracting aging. An examination of polymorphisms and gene expression levels in human DNA repair genes and a comparison with the equivalent genes in shorter lived mammalian species should help determine the importance of DNA repair in normal aging processes.

Human Transcriptome Map

The chromosomal position of 45,049 human expressed sequence tags, (ESTs) and genes belonging to 24,106 UniGene clusters are given by GeneMap'99. To obtain an expression profile of these genes, HGP made use of the SAGE technology and databases. SAGE can quantitatively identify all transcripts expressed in a tissue or cell line. It is based on the extraction of a 10-base pair (bp) tag from a fixed position in each transcript and the sequencing of thousands of these tags. Software programs and databases support the identification of the mRNAs corresponding to the tags in a SAGE library. However, this step is prone to errors, and tag assignment requires manual verification.

The National Centre for Biotechnology Information (NCBI) SAGEmap database has electronically extracted tags from mRNAs and ESTs in UniGene clusters. A manual check of 156 tags extracted from 30 UniGene clusters showed that wrong tags mainly stemmed from sequence errors in ESTs and from errors in their 5′ and 3′ orientations.

HGP developed algorithms to select 3′-end clones of 713.489 ESTs assigned to UniGene clusters and identified their tags.

Sequence comparison algorithms discarded tags caused by sequence errors while preserving tags from alternative transcripts or single nucleotide polymorphisms.

HGP identified reliable tags for 18,954 of the 24,106 UniGene clusters mapped on GeneMap'99. The manual analysis of 287 tags extracted from 86 UniGene clusters from intervals of chromosomes 1 and 22 showed an error rate of 6.2 per cent in electronic tag identification algorithms. To check for errors in UniGene clustering, HGP verified tags on the available sequenced PI-derived artificial chromosomes (PACs) of the mapped markers and annotated them accordingly.

Human Transcriptome Map uses these tag assignments to relate 2.31 million tags in public SAGE libraries and 160,000 tags in HGP neuroblastoma SAGE libraries to the UniGene clusters mapped in GeneMap'99. The Human Transcriptome Map shows expression profiles for any chromosomal region in 12 tissue types. SAGE libraries of a specific tissue were combined into tissue-specific libraries. HGP included tissues for which 100.000 or more tags were available, as most transcripts in a tissue are represented in a library of this size. Five libraries represent normal tissues and seven libraries represent tumour tissues.

The Human Transcriptome Map has three levels of resolution. The whole chromosome view shows gene expression per chromosome. Each horizontal blue or red bar represents the expression level of a UniGene cluster. UniGene clusters mapped by several markers are shown only once, at the position of the highest reliability. The identity, map position, and precise expression of the genes are shown in the concise interval view. The highest resolution is given by the extended interval view, where expression levels are shown for all individual tags of a gene.

Whole chromosome views reveal a higher order organisation of the genome, as there is a strong clustering of highly expressed genes. Chromosome 11 has several large regions of high gene expression, interspersed with regions where gene expression is low. This pattern is observed in all 12 tissues. An application of a moving median with a window size of 39 genes to the chromosome 11 map even more clearly visualises the expression differences.

Most chromosomes show these clusters of highly expressed genes, which is known as RIDGEs. A quantitative definition of RIDGEs is not straightforward, as there is a continuum from small to very large clusters. HGP analysed whether RIDGEs can be explained by a random variation in the distribution of highly expressed genes among the 18,954 genes of the Human Transcriptome Map. When defined as regions in which 10 consecutive moving medians have a lower limit of four times the genomic median, HGP identified 27 RIDGEs. The probability of observing this number of RIDGEs under a random permutation of the order of the 18,954 genes is very low.

In addition, Bayesian statistical modelling without prior cluster definition showed that a model of non-random distribution provided the best fit with the observed clustering. These analyses show that RIDGEs most likely represent a higher order structure in the genome. Many RIDGEs have a high gene density. Chromosome 18 is, on average weakly expressed, and only 385 genes have been mapped to it on GeneMap'99. The equally large chromosome 19 consists of a succession of RIDGEs and harbours 937 mapped genes.

Although many human genes are still unmapped, the difference in gene density of chromosomes 18 and 19 is supported by CpG island density analyses. The correlation between RIDGEs and gene density is even more suggestive for chromosomes 3 and 6. The RIDGE on chromosome 6 corresponds to the major histocompatibility complex (MHC) region.

A correlation between gene expression and density of mapped genes is found for 50 to 60 per cent of the RIDGEs. Typical RIDGEs count six to 30 mapped genes per centiray, compared to one to two mapped genes per centiray for weakly transcribed regions. In RIDGEs average expression levels per gene are up to seven times that of the genomic average. This suggests that in RIDGEs, transcription per unit length of DNA is 20 to 200 times that in weakly expressed regions.

About 40 to 50 per cent of the RIDGEs are not gene dense. These RIDGEs preferentially map to telomeres, which is remarkable in light of the observed telomeric silencing in yeast. Chromosomes 4, 13, 18 and 21 show an overall low gene expression and are devoid of RIDGEs. The latter three chromosomes are responsible for most constitutional trisomies, suggesting that the low expression and low gene density could limit the lethality of an extra copy of them.

The Human Transcriptome Map provides a tool to identify candidate genes that are over-expressed or silenced in cancer tissue. Neuroblastomas frequently show amplification of the distal chromosome 2p region, which targets the *N-myc* oncogene. A comparison of the whole chromosome views of chromosome 2p shows over-expression of two adjacent genes in neuroblastoma SAGE libraries.

The extended interval view identifies these genes as N-myc and the often coamplified neighbouring gene DDX-1. Therefore, global positional information of chromosomal defects is sufficient to identify candidate oncogenes. Also, tumour-specific down-regulation can be detected. Examples are a cluster of five matrix metalloproteinases on chromosome 11 that are down-regulated in breast cancer tissue; the E-cadherin tumour suppressor gene on chromosome 16 (406 cR) that is down-regulated in breast cancer tissue, as compared to normal breast tissue; and five carcinoembryonic antigen-related cell adhesion molecule genes on chromosome 19 that are down-regulated in colon carcinoma tissue, as compared to normal colon tissue.

Error sources in the Human Transcriptome Map are clustering errors in UniGene and the assignment of wrong tags to UniGene clusters. HGP's algorithms assign ~6.2 per cent erroneous tags to UniGene clusters. The influence of these errors is probably attenuated. Assuming a total of 100,000 genes with two tags each, 200,000 tags would represent all human genes.

Since there are >1 million variants of a 10-bp tag sequence, -80 per cent of the erroneously extracted tags will not match tags present in SAGE libraries and therefore will not influence overall expression profiles. However, individual tags and expression levels of UniGene clusters may harbour errors and require experimental confirmation.

To test whether errors in UniGene clustering and mapping to GeneMap'99 may influence HGP's observation of RIDGEs, HGP constructed a sequence-based expression map for the annotated chromosome 21 sequence and for a 4.3-Mb annotated contig of the MHC region on chromosome 6. These maps showed that the MHC region is a pronounced RIDGE, whereas chromosome 21 is devoid of RIDGEs and has an overall weak gene expression. Therefore, the higher order structure of the genome observed with the Human Transcriptome Map will largely be correct.

The existence of RIDGEs is unanticipated, as a comparable SAGE-based transcriptome map for yeast showed an even distribution over the genome of highly and weakly expressed genes. Since the Human Transcriptome Map identifies different types of transcription domains, it can now be analysed as to how they relate to known nuclear substructures, such as nuclear speckles, PML bodies, and coiled bodies. A definition of the position of tags to the full chromosomal sequences will further increase the resolution of the transcriptome map.

Chimeric Genes in the Hominidae Lineage

Due to their recent history, two human chimeric genes, *PMCHL*1 and *PMCHL*2, open an unprecedented way to analyse the molecular mechanisms of gene remodeling and selection of functions that have operated during the late stages of primate evolution. The *PMCHL* genes were named pro-*MCH*-like 1 and 2 genes (*PMCHL*1 and *PMCHL*2) on the basis of partial identity to the *MCH* gene. The human *MCH* gene maps on chromosome 12q23 and encodes a neuropeptide precursor, whereas *PMCHL1* and *PMCHL2* are located onto human chromosome 5p14 and 5q13, respectively, and correspond to 5′–end truncated versions of the *MCH* gene.

PMCHL genes arose recently during primate evolution by a first event of truncation/transposition from the ancestral chromosome 12 to the ancestral chromosome 5p about 25 to 30 Ma, i.e., before divergence of the *Cercopithecoidae*. This was followed by a second duplication event, which operated in the *Hominidae* lineage about 5 to 10 Ma and which distributed the two genes on each side of the chromosome 5 centromere.

Both unspliced sense and antisense transcripts from the *PMCHL1* gene but not the *PMCHL2* gene have been observed in different areas of the developing human brain. A puzzling issue concerns the relation between their recent emergence and their putative function or, more precisely, whether the *PMCHL* genes are functional genes not previously characterised or inactive pseudogenes. Focus on the molecular mechanisms responsible for the emergence of *MCH-derived* sequences onto human chromosome 5 had first come from parallel studies on the regulation of *MCH* gene expression undertaken in the laboratory. Recently in human and rodents, HGP showed two classes of antisense RNAs complementary to the *MCH* gene:

i) spliced-variant mRNAs complementary in their 3′ end to the *MCH* gene, encoding newly found DNA/RNA binding proteins, and
ii) short noncoding unspliced PNAs that overlap only the coding part of the *MCH* gene and initiate at cap site CS3-5. This transcriptional unit was named *AROM* for antisense-RNA-over-*lapping-MCH gene.*

An analysis of the structure of the *PMCHL* genes revealed the presence of a stretch of A at the end of the MCH-derived portion that exactly coincidates with one of the polyadenylation [poly(A)] sites found within the *AROM* gene, polyA(b). This led to the conclusion that a MCH-derived sequence likely was inserted in the ancestral chromosome 5p by an event of retrotransposition of an *AROM* messenger RNA, incidentally strongly expressed in testis.

By combining *in silico* screening and direct sequencing of bacterial artificial chromosome (BAC) clones specific to the chromosomal regions 5p14 and 5q13, the genomic structure of the *PMCHL* genes was further compared. According to the Web survey, several expressed sequence tags (ESTs) were found in two categories:

i) 3′ cDNA clone IMAGE ah92fl.l.sl and qf54b04.xl, which are parts of *PMCHLI* spliced sense transcripts, and

ii) 3′ cDNA clone IMAGE qf66aO4.xl, al54h4.sl, and al47h07.sl, corresponding to parts of *PMCHL2* unspliced antisense transcripts and indicating that the regulation of the expression of the *PMCHL* genes was far more complex than previously thought.

Structural analysis of those genes was refined by using rapid amplification of cDNA ends and polymerase chain reaction (RACE-PCR) and reverse transcriptase-PCR (RT-PCR) in conjunction with the genomic analysis.

HGP revealed *PMCHL1* / *PMCHL2* gene expression in human testis and established the precise 5′ and 3′ ends of the sense and antisense *PMCHLI* RNA unspliced products previously described in different areas of the human brain. HGP also found in human foetal brain and in human adult testis several classes of alternative spliced mRNAs.

On both loci, MCH-derived, retrotransposed sequences recruited a group of downstream exons and introns into their transcription units thereby creating previously unknown genes with a chimeric structure. The existence of such an impressive variety of *PMCHLI* and *PMCHL2* transcripts resulted from the use of four polyadenylation sites (Al-A4) and a tissue-specific modulation of alternative splicing.

Several cap sites were also found on the basis of RACE-PCR experiments. *PMCHL2* cap sites were mainly located from 500 base pairs (bp) to more than two kb upstream to the insertion site, whereas *PMCHLI* cap sites were found 500-bp upstream as well as 50- to 100-bp downstream to the insertion site.

However, because of the complex population of mRNAs in all the tissues analysed, it was not possible to assign a precise cap site to each class of mRNAs. Even though HGP cannot exclude artifactual pausing of the reverse transcriptase during synthesis of the cDNA products, this suggests that alternative splicing coupled to different starting points of transcription is probably a mechanism that allows the cell to generate a wide repertoire of *PMCHL* genes transcripts.

The longest open reading frames (ORFs) initiated from an ATG codon in a reasonable translation initiation context were deduced from the mRNA sequences obtained by RACE-PCR and RT-PCR. Two major classes of ORFs ($\geq$33 amino acids) were found regardless of the alternative splicing pattern:

i) ORFs encoded by exon I and intron A (unspliced RNAs) and exon 1/exon 2/exon 2′ (spliced RNAs) exhibit a strong similarity with pro-MCH, and

ii) ORFs encoded by exons 4 to 5a and 5b display no sequence similarity with known proteins. No ORF of large length could be found for antisense RNAs.

Sense unspliced *PMCHLI* transcripts may produce a nuclear localisation signal (NLS)-containing protein deduced from ORF 1 sequence in an *in vitro* translation assay and in transfected Cos cells. Direct proofs of the translational ability of the spliced mRNA products described here are still lacking.

However, that both *PMCHL1* and *PMCHL2* are specifically and differentially regulated in testis and that only *PMCHLI* is expressed in human fetal as well as newborn and adult brains is consistent with the conclusion that those newly originated genetic elements are transcriptionally active and tightly regulated genes. To determine whether the divergent expression patterns of *PMCHLI* and *PMCHL2* could be explained by a different genomic environment in the flanking regions, HGP expanded comparative analysis of the genomic structure of the *PMCHL* genes.

The nucleotide sequence of the *PMCHLI* and *PMCHL2* genomic regions over 17-kb revealed similar genomic environments with strong sequence identity (981/6) between the 5p14 and 5ql3 loci. To further delineate the extent of the region duplicated on both arms of chromosome 5, HGP performed fluorescent *in situ* hybridisation (FISH) analysis on human metaphase chromosomes with several BAC clones bearing the *PMCHLI* locus and extending more than 100-kb both 5′ and 3′ to this gene. All those clones displayed the same hybridisation patterns with strong cross-hybridisation on both arms of the human chromosome 5 at bands 5p14 and 5q13.

The event of duplication that took place 5 to 10 Ma involved a large region of ancestral 5p14 encompassing several hundreds of kilobases. However, further studies are required to delineate the particular environment of cis-regulatory elements driving the striking tissue specific expression of both *PMCHL* genes.

The source of the 5′ exons was identified as a retrotransposed sequence originated from the *MCH*/*AROM* locus. However, the origin of 3′ exons remained unclear. HGP examined several hypotheses concerning the origin of these non-MCH-derived *PMCHL* exons:

i) these exons might be part or duplicate of an unrelated previously existing gene, supporting the concept of exon shuffling or, alternatively,
ii) these exons might originate from a unique genomic sequence that fortuitously evolved as a standard intron-exon structure and regulatory sequences for *PMCHL*.

To study the early molecular evolution of the *PMCHL* transcription units, HGP first performed a FISH analysis on mouse metaphase chromosomes with BAC clones surrounding the area of insertion of the *MCH-derived* sequences (namely, 2303C18, 34413, and 811M22). Only the 811M22 BAC clone, bearing the 3′ *PMCHL* exons but not the 5′-transposed portion of the gene, displayed a clear unique hybridisation signal. This signal was found onto the pericentromeric region of the mouse chromosome 15.

After comparing this result with the mapping data found in the 'Mendelian Inheritance of Man gene map' and mouse to 'human homology region map' databases, the transposed *MCH* sequence was inserted in a region close to the site of evolutionary rearrangement that disrupted the conserved synteny relationship with the mouse *Mus musculus* genome from MMU13 to MMU15. Furthermore, probes bearing the 3′ exons did not reveal cross-hybridisation signal on mouse and primates and these exonic sequences did not display any similarity to any sequence of the GenBank database except the IMAGE cDNA clones. This ruled out the hypothesis that the 3′ exons might be a duplicate of an unrelated previously existing gene.

However, this does not exclude that the retrotransposed sequence may have been inserted in a pre-existing gene on 5p. To test this alternative, the phylogeny of the *PMCHL* intronic and exonic sequences was analysed. HGP attempted to amplify the corresponding region from DNA samples from nine species of primates and from mouse by using the set of primers used to amplify intronic and exonic sequences of human genomic DNA.

Several PCR products of the same size as those obtained from human DNA were obtained from seven primate species [*Pan troglodytes* (PTR), *Pan paniscus* (PAN), *Pongo pygmaeus* (PPY), *Hylobates lar* (HLA), *Cercopithecus hamlyni* (CHA), *Papio papio* (PAP), *Cebus capucinus* (CCA)]. All of the amplified products obtained from anthropoids were sequenced and compared with the human DNA sequence.

Comparative phylogenetic analysis of the *PMCHL* intron-exon boundaries revealed that consensus sequences at the 5′ donor splice site and in the 3′ acceptor splice site of the *PMCHL1* intron A were conserved in all the primates, suggesting existence of a functional constraint. Similarly, strong conservation of sequences was noted at the intron B and C boundaries.

In contrast, a splice donor site in intron D was created in *Cercopithecoidae* (CHA) as a result of a C to T substitution at nucleotide +2. Alternative splice acceptor sites for exon 5a and exon 5b were also created by nucleotide substitution, GA to AG in *Hylobatidae* (HLA) and G to A at nucleotide +1 in *Cercopithecoidae* (PAP and CHA), respectively. Poly(A) signals PS2 and PS3 corresponding to the poly(A) addition sites A2 and A3 were also found to be the sites of mutations. Interestingly, a C nucleotide was found at nucleotide +3 of PS2 in CHA and PAP but not CCA, suggesting that this mutation arose specifically in the *Cercopithecidae*.

Furthermore, HLA possess the same ATTAAA sequences as the ones found in human, whereas CCA, PAP, and CHA have GA and TC at nucleotides + 3 and + 4 in PS3. Therefore, these results are consistent with the hypothesis that the 3′ art of the *PMCHL* transcription unit evolved from non-coding DNA in a common ancestor of Hominoids as a result of the creation of standard intron-exon boundaries and poly(A) signals that have been conserved in humans. In CHA and PAP, which do not carry functional splice sites, HGP succeeded in amplifying only a small part of *AROM/MCH* retrotransposed sequence from the genomic DNA. In addition, a strong divergence of *PMCHLI* sequence was noted in these species reflecting weak selective constraint.

The similar exon structure of the *PMCHL* genes found in HSA, PAN, PTR, PPY, and HLA together with the divergence of sequence of the retrotransposed *AROM/MCH* sequences in the *Cercopithecoidae* indicates that there was a relatively short time between the first insertion event and the subsequent mutation events leading to the recruitment of intronic and exonic components into a functional transcription unit and the speciation.

As expected for emerging functions, the underlying genes were likely to undergo fast divergence until they gained stronger physiological constraints. This strongly suggests that the *PMCHL* gene was conserved in *Hominidae* due to the acquisition of some constraints, probably an emerging role in primates.

The results of HGP reveal the molecular, genetic, and evolutionary mechanisms that participated in the origin of two chimeric functional genes *PMCHLI* and *PMCHL2* in the *Hominidae* lineage. Taken together, the data on the tissue-specific expression and the conserved features of the *PMCHL* genes suggest that their mRNA or protein have been exapted into a functional role in the primate lineage.

The identification of the many processes in genome evolution have shown that *de novo* generation of building blocks—single genes or gene segments coding for protein domains—seems to be rare. Instead, gencime novelty was mainly built by modification, duplication, and functional changes of the available blocks by processes of gene duplication, exon shuffling, or retrotransposition of genes.

In the context of human genome evolution, the previously unknown mechanism of transcript fusion of the adjacent *Kua* and *Uev* genes was recently proposed to create a chimeric *Kua-Uev* mRNA and the cognate fused protein. However, in the case HGP described the recruited portion fused to the AROM/ MCH-derived sequences was shown to have originated from a unique noncoding sequence. Moreover, the complex structure and evolutionary history *of PMCHL* encompass several phenomena pointing to an important role for introns in the origin *of* newly characterised genes, as the exon theory of gene has suggested:

i) emergence of the 5′ exons by an event of duplication *of* a 5′-end truncated part of the *MCH* gene via a process of retrotransposition of an antisense *MCH* mRNA;

ii) creation of 3′ exons from a unique noncoding genomic sequence that fortuitously evolved as a standard intron-exon structure and polyadenylation signal sequences;

iii) alternative transcriptional initiation and splicing processes, further complicated by the presence of antisense RNAs; and

iv) a nested gene encoding unspliced mRNAs products.

In the context of genome research, the existence of such gene structures poses a particular dilemna in the perspectives of prediction of exons from genome sequence data. In fact, the complex gene structure of the *PMCHL* loci, was not predicted from the genome sequence and exon prediction programmes.

High-Resolution Radiation Hybrid Map

Over the past several years, an international consortium of scientists has taken advantage of technical improvements in DNA sequencing and mapping technologies to generate a working draft of the human genome sequence. The first step in this effort involved the construction of bacterial artificial chromosomes (BACs), each containing a stable segment of approximately 160 kilobase pairs (kbp) of the human genome. Collections of human BACs estimated to represent more than a 10-fold redundancy of the human genome were used to generate BAC maps. This process resulted in a minimally redundant set of BACs, assembled into physically separate contigs, representing the majority of the human genome and serving as the substrate for large-scale DNA sequencing.

The human DNA segment of each BAC was fragmented into a redundant collection of smaller clones, which were each sequenced and assembled into a limited number of sequence contigs per BAC. All DNA sequence within each physical contig consisting of multiple BACs was assembled.

The BAC contigs used as substrates for sequencing are separated by physical gaps of unknown size, and given that the multiple sequence contigs representing each single BAC insert are often unordered and unoriented with respect to each other, the sequence product is referred to as the working draft to distinguish it from completed genomic sequence in which the size of all gaps and the order and orientation of all sequence is known.

HGP has used the technique of radiation hybrid (RH) mapping, in conjunction with more than 40,000 unique human STSs, to identify those segments of the human genome that are absent from the working draft sequence, to provide independent estimates of the size and location of missing sequence in relation to the existing sequence.

Construction of human genome topography has been greatly facilitated in recent years by the development of STSs as genomic landmarks. Each STS is defined by a short segment of 100 to 500 base pairs (bp) of human DNA sequence and is assayed by means of polymerase chain reaction (PCR). Common sets of such sequence-based markers can be easily screened and therefore can be used to integrate maps constructed by different mapping methods.

The development of STSs from short DNA sequences derived from a variety of clone sources provides a method to obtain markers from regions of the human genome that may be difficult to clone by any single vector system. Recent experience suggests that up to 10 per cent of certain gene-rich regions of human chromosome 21 are composed of such 'hard-to-clone' DNA.

HGP have developed a large set of STSs using DNA sequence derived from a diverse set of sources in an effort to maximise the coverage of the human genome. The strategy involved an initial electronic analysis of genomic DNA sequence to eliminate repetitive DNA sequences, followed by an automated selection of oligonucleotide primers to generate PCR products 90 to 350 bp in length under a single set of reaction conditions. PCR products were assayed by ethidium bromide staining after agarose gel electrophoresis. An STS was judged successful when the primers produced a distinct PCR product of the expected size from total human DNA and failed to produce a product of this size from either hamster or mouse genomic DNA.

HGP generated a total of 41,234 human STSs that met these criteria. Of these STSs, 14,953 were scored on rodent-human hybrid somatic cell mapping panels to determine their chromosomal location. A total of 14,041 of these 14,953 STSs could be assigned to a unique human chromosome. These 14,041 chromosome-specific STSs, as well as the remaining 26,281 STSs not scored on the chromosomal mapping panel, were used to construct a high-resolution RH map of the human genome.

Random fragments of human chromosomes are retained in somatic cell hybrids between irradiated diploid human cells and nonirradiated hamster cells in the absence of selection for human chromosomal material provides the basis for RH mapping. DNA isolated from a panel of 80 to 100 such independent RH clones serves as a mapping reagent for ordering STSs and for determining the distances between them in the human genome. In this approach, the frequency of irradiation-induced breakage between two human markers is used as a measure of distance, and marker order is determined in a manner analogous to meiotic linkage mapping.

As in the case of meiotic linkage mapping, the relative confidence in alternative marker orders in RH maps can be assessed by comparing LOD scores (logarithm of the odds ratio of linkage versus no linkage) using appropriate maximum likelihood statistical methods.

An important advantage of RH mapping is that hybrid panels for constructing maps at very different levels of resolution can be generated by experimentally manipulating the dose of irradiation. RH maps of the human genome published to date have used either the G3 RH panel, which was constructed by using 3000 rad of x-rays, or the G3 RH panel, which was constructed by using 10,000 rad of x-rays. Both of these RH panels provide RH maps with good long-range continuity.

However, STSs separated by less than 1, 000 kbp in the genome are not ordered routinely with high confidence when these mapping panels are used, due to the relatively small number of chromosome breaks. In contrast, a panel of 90 independent RH clones constructed at the Stanford Human Genome Center (SHGC) with 50,000 rad of x-rays (the TNG panel) allows STSs separated by less than 100 kb in the genome to be ordered with high confidence. The price of this increased resolution is that a large number of STSs need to be scored on the TNG panel to produce RH maps with good long-range contiguity.

All the 40,322 STSs were scored in duplicate on the each of the 90 clones of the TNG panel. A subset of 10,227 of these STSs were also scored in duplicate on the 83 clones of the G3 RH panel. HGP used these TNG and G3 RH data in conjunction with the chromosomal assignments of the 14,041 STSs, to assign an additional 25,486 STSs to unique human chromosomes. All 39,527 STSs that could be assigned to a specific chromosome were used for RH map construction.

Chromosomal assignment of STSs before map building allowed HGP to construct the RH map for each chromosome independently, rather than using all data simultaneously. This approach greatly reduces the number of pairwise comparisons required in map construction and results in improved power to determine marker order. G3 RH data for a given chromosome were used to construct a contiguous RH map. In total, 8, 351 of the 10, 227 STSs scored on the G3 panel (82 per cent) were ordered with respect to each other on this G3 RH map of the human genome, with an average physical distance between ordered markers of 381 kbp. Only 275 of the 8, 328 LOD scores between adjacent STSs on the map were less than 3.0.

The segments of the genome flanked by two low LOD scores were ordered and oriented by means ofSTSs on the G3 map that were also included in the Genethon meiotic linkage map and which provided order information based on that map. Markers separated by less than a physical distance of 1, 000 kbp were not ordered routinely with high confidence. An additional 1466 unique STSs, each with an RH vector identical to an ordered marker, were assigned the same map position as the ordered STS, placing a total of 9, 817 STSs on the G3 map.

HGP used the TNG data to build chromosome-specific maps. In total, 33,627 STSs of the 39,527 STSs used in TNG map construction (85 per cent) were ordered with respect to each other on the high-resolution TNG map of the human genome, with an average physical distance between ordered markers of 94 kbp. Only 3, 676 of the 33, 604 LOD scores between adjacent STSs on the TNG map (11 per cent) were less than LOD 3.0.

HGP defined an STS contig as a segment of the TNG map flanked by two LOD scores less than 3.0. G3 map position was used to help order and orient the STS contigs on the TNG map relative to one another, because 8, 961 of the STSs with map locations on the G3 map were also placed on the TNG map. An additional 3, 051 unique STSs, each with an RH vector identical to an ordered TNG marker, were assigned the same map position as the ordered STS, giving a total of 36, 678 STSs on the TNG map out of a total of 39, 527 STSs used in map construction (93 per cent).

RH mapping provides distance measures between adjacent markers in the map. This distance, based on the frequency of breakage between two markers in the radiation hybrid clones, is measured in units called centirays (cR). HGP compared cR distances between STSs on the TNG and G3 maps with actual physical distances between these STSs based on DNA sequence and determined that 1 cR on the TNG map corresponds to an average of two kbp of physical distance, whereas 1 cR on the G3 map corresponds to a physical distance of 13 kbp.

A comparison of 1, 778 pairs of STSs where each member of a pair had an identical RH vector and thus a distance of 0 cR, revealed that in 86 per cent of cases, the STSs defining the pair were separated by less than 20 kbp of physical distance. Distances less than 10 cR were found to overestimate the true physical distance, while distances greater than 70 cR were found to underestimate the true physical distance. Of the 33, 604 cR distances between adjacent STSs on the TNG map, 30, 420 (91 per cent) are less than 100 cR, equal to a physical distance of less than 200 kbp.

In 512 adjacent STS intervals, the STSs appeared to be completely unlinked, and HGP were unable to calculate a cR distance. In these instances, HGP picked STSs flanking the gaps that had also been ordered on the G3 map and used the G3 map distance information to estimate the physical distance of the gap.

Before the availability of a large amount of draft sequence, the only method for determining if a specific STS marker mapped to a specific DNA clone was to carry out a PCR reaction with DNA from the clone of interest. Although simple in principle determining which of tens of thousands of STSs are present in each of tens of thousands of clones is a formidable and expensive task.

In contrast, with the availability of the draft sequence, one can map *in silico* (through computer analysis) by electronically comparing the DNA sequence of an STS of interest to all of the available draft sequences. In this way, it is possible to identify which clones contain the STS of interest and determine the precise location of the STS on each clone.

HGP used an *in silico* mapping approach to determine which of the STSs present in the TNG RH map are also present in the working draft sequence. Due to one of the primary goals in this work was to determine which, if any, STSs in the RH map were absent from the draft sequence, HGP needed a method with high sensitivity. Although 'electronic PCR' can be performed with only oligonucleotide primer information and the known size of the STS sequence, this method identifies only 80 per cent of STSs known to be present in a given sequence contig, even allowing for single-base differences between the primer sequence and the draft sequence.

Comparing the entire sequence of the STS with the draft sequence is a more sensitive method of in silico mapping.

However, this approach is complicated by repeated sequences that are present internal to the primer sequences of an STS, which pose no problem for specificity when performing a PCR reaction, but which can lead to significant false positive hits when sequence alignment algorithms are used.

Although it is difficult to rigorously determine the ideal in silico mapping parameters, HGP used the BLAST algorithm, requiring an alignment of 100 bases or greater with an identity match of 90 per cent or greater and an E value of less than one. $0E^{-45}$, to compare STS sequences with the working draft sequence. This provides a good compromise between specificity and sensitivity for this electronic mapping approach. It was found that only 15 of the 971 STSs (1 per cent) ordered on the TNG map of chromosome 21 failed to hit the finished DNA sequence of chromosome 21 when these *in silico* conditions were used.

HGP compared all 36, 678 STSs on the TNG RH map with the 5 September 2, 000 unassembled GenBank human sequence data release, using the in silico mapping parameters described above. Only 3, 604 of these STSs fail to hit a GenBank sequence, suggesting that the publicly available sequence covers more than 90 per cent of the human genome. Due to the finished state of the chromosome sequence, the chromosome with the lowest percentage of nonhit STSs is chromosome 21 with one per cent missing. Next lowest are chromosomes 20 and 14, with only two per cent and three per cent of STSs missing, respectively. Chromosomes 16 and 8 have the highest percentage of STSs with no sequence hits, with 15 per cent and 14 per cent missing, respectively.

The sequence coverage of the genome appears to be fairly uniform. Cases in which STSs exist that have no counterparts in the sequence database serve as valuable reagents for completing the sequence of the human genome. STSs that fail to hit a match in the presently available sequence can be used to screen a variety of human DNA libraries cloned in different vectors, providing an efficient method for closing existing clone gaps in the human draft sequence. By assigning each STS that hit a sequence in the working draft to a specific sequence, the draft sequence can be linked to the RH map. This provides an independent measure of order for the clones that were used to generate the draft sequence of the genome and an estimate of the physical length of gaps between non-overlapping clones.

On numerous occasions, a single STS was found to hit multiple sequence accessions with a variety of different E values, and in each case, meeting the minimal criteria for a hit described above. In such cases, HGP considered only those hits with the most significant E value for further evaluation. Even with this criterion imposed, the 33, 078 individual STSs on the TNG map with at least one hit in the draft sequence produced a total of 54, 225 hits, giving an average of 1.7 hits per STS.

For every STS with more than one accession hit, HGP chose the accession with the greatest number of hits by other STSs mapping to the same chromosome and linked the STS to this unique accession. In this way, HGP created a minimal set of 15, 718 accessions that contained a single hit for each of the 33, 078 TNG map STSs. Of this minimal accession set, 7, 957 contained more than one STS hit, providing an opportunity to determine how often multiple STSs hitting a single clone had discrepant chromosome assignments.

HGP found that 475 of these 7, 957 accessions, contained STSs that mapped to two or more chromosomes. At this time, it is unclear if these discrepancies are due to errors in chromosomal assignment, if they represent false positive electronic mapping hits, or if they represent chimeric clones that actually contain human DNA from more than one human chromosome.

The primary substrate used to generate the draft sequence of the human genome as a minimally redundant set of BAC clones assembled into physically separate clone contigs by restriction fingerprint mapping. Although fingerprint mapping has the advantage of mapping the clones that are used for DNA sequencing, it has the disadvantage that the restriction fragments used to build restriction maps are not sequence-based like the STS markers used in RH mapping.

HGP compared the order of sequence accessions in the TNG map with the 5 September 2000 data release of ordered accessions, which is posted on the Washington University Genome Sequencing Center website This data set consists of 31,102 ordered sequence accessions grouped by fingerprint mapping into 1001 physically distinct clone contigs. HGP determined that 14, 363 of the 15,718 sequence accessions on the TNG map (92 per cent) are included in the fingerprint data set and provide a basis for comparison of the two maps. A total of 773 of these 14,363 accessions (5 per cent) were assigned to a different chromosome in the TNG map as compared to the fingerprint map.

HGP compared the order of 8, 976 sequence accessions with the same chromosome assignments in the fingerprint and TNG maps in the following manner. The accessions for each chromosome were ordered on the basis of the fingerprint data and numbered sequentially, beginning with one, providing each accession on the chromosome with a unique fingerprint order number.

All sequence accessions on the chromosome then received a TNG order number, reflecting its position in the TNG chromosome map. The list was then sorted by fingerprint order number, and the pattern of TNG order numbers was analysed. Moving from top to bottom on the sorted list, HGP calculated the absolute difference in TNG order number for each pair of adjacent STSs, and then determined the number of all intervals differing by one position, two positions, three positions, and so forth.

If the fingerprint order and the TNG order were completely consistent, all intervals defined by adjacent STSs would differ by one position. This method of comparing the orders of shared markers in two different maps has high practical value, because it provides information regarding the number of order discrepancies as well as the magnitude of those discrepancies.

The results of the comparison between the TNG map and fingerprint map revealed that of 8, 975 total intervals, 7, 722 (86 per cent) differed by five positions or fewer, whereas 856 intervals (10 per cent) differed by more than 10 positions. Although this analysis provides a direct comparison of sequence accession order in the TNG map versus the fingerprint map, it does not provide information regarding which map, if any, most often reflects the true order of sequence in the human genome. To address this question, HGP carried out a similar analysis comparing the TNG map and the fingerprint map to the assembled draft sequence of the human genome.

Sequence information derived from all STSs on the TNG map was used by David Haussler and colleagues to perform electronic PCR with the assembled draft sequence and determine the order of the STSs in this assembly. Given that both the fingerprint map and the TNG map were used in conjunction with other mapping information to assemble the draft sequence, comparison of the TNG and fingerprint maps with the draft sequence does not provide truly independent validation. Nevertheless, such a comparison provides a useful means of examining each map in relation to the final draft sequence product. Of 8, 975 total intervals, 7, 744 (86 per cent) differed by three or fewer positions when the TNG order was compared to the draft sequence order. In contrast, 8, 227 of the same 8, 975 total intervals (92 per cent) differed by three or fewer positions when the fingerprint map order was compared to the draft sequence order.

Although both maps have inconsistencies with the assembled draft sequence, the discrepancies between the draft sequence and map order are often at different positions in the TNG and fingerprint maps. Thus, using the order information from both of these maps as well as others to produce the final assembly of the draft sequence has resulted in a much better quality product than would have been produced by using either map alone.

The fluorescence in situ hybridisation (FISH) is another important physical map, which has provided order information for the draft sequence. The TNG map contains 886 unique STSs derived from clones, which have been mapped by FISH. Of these 886 markers, 28 (3 per cent) have a different chromosomal assignment on the FISH map compared to the TNG map. An analysis of the order of the 858 markers that have a consistent chromosome assignment on both maps reveals that 836 (97 per cent) have a consistent order on both the TNG and FISH maps.

Draft sequence of the human genome is a tremendous achievement, which will enable scientific discovery in ways not previously possible. However, far greater scientific value will be gained from the completed sequence of the human genome. In addition to DNA sequencing efforts of the International Consortium, Cetera Genomics has used a whole-genome shotgun approach to produce an assembled sequence of the human genome.

In an effort to determine how the Cetera sequence adds information to the working draft sequence and vice versa, Mark Adams and Peter Li at Cetera have used sequence alignment methods to assign 34,725 SHGC STS sequences to unique positions in the Cetera sequence and provided HGP with the ordered list of STSs. HGP compared this list with the 33, 627 SHGC STSs ordered on the TNG map and with the 27, 471 SHGC STSs assigned to unique locations in the 7 October freeze final assembled draft sequence by David Haussler and colleagues.

A total of 20, 874 STSs are shared in common in the three lists. Because each group used different methods to identify STSs in the sequence, it is not appropriate to use the STS list to draw any conclusions regarding coverage of the genome. However, the 20, 874 STSs spaced at an average distance of 150 kbp in the genome provide valid information for assessing the consistency of STS chromosome assignment and order in the draft sequence, the Cetera sequence, and the TNG map. Of the

20, 874 STSs, only 766 (3.7 per cent) revealed a chromosome discrepancy.

In 220 of these cases (29 per cent), the TNG chromosome assignment was different from that shared in both the draft and Cetera sequences. In 331 cases (43 per cent), the Cetera chromosome assignment differed from that shared by the draft sequence and the TNG map. In 69 cases (9 per cent), the chromosome assignment in the draft sequence differed from that shared by the Cetera sequence and the TNG map.

In 146 cases (19 per cent), the draft sequence, the Cetera sequence, and the TNG map each had a different chromosome assignment for the STS. Given the caveat that the data sets are not completely independent, those chromosome assignments supported by two of the three data sets are most likely the correct assignments. These results illustrate the utility of comparing multiple data sets to identify potential errors.

For each STS, HGP identified the STS that was immediately distal in the draft sequence order and determined if the two STS were adjacent in the same orientation in the Cetera sequence. Of the 20, 006 comparisons, 12,791 (64 per cent) revealed an identical order and 8, 183 (36 per cent) revealed a discrepant order. The result is the same if HGP begin with the STSs in the Cetera sequence order rather than the draft sequence order.

Of the discrepancies, 51 per cent represent pairs of markers that are inverted in the draft sequence versus the Cetera sequence, and 86 per cent of all discrepancies are separated by five positions or fewer in the draft sequence versus the Cetera sequence. Comparison of the order of discrepant STSs with their order in the TNG map revealed that the TNG order was consistent with one of the two sequence orders -60 per cent of the time. In these cases, the ING order was consistent with the Cetera order slightly more often than the draft sequence order.

For any given discrepancy, if the TNG map was found to be consistent with one of the two sequence orders, it was nearly as likely to be the draft sequence order as the Cetera sequence order. These results again illustrate the utility of using multiple data sets in an effort to resolve discrepancies. Resolution of discrepancies will play an increasingly important role in ongoing efforts to complete the sequence of the human genome.

Although the path to produce finished sequence of the human inserts in each BAC clone is well defined, the approach for identifying and isolating those sequences missing from the available sequence is less clear. The 3, 604 STSs ordered on the TNG map, which are not present in the draft sequence, serve as very useful reagents for isolating the missing portions of the human genome. Of these 3, 604 STSs, 2, 472 (69 per cent) are present on the 'ist of SHGC STSs identified in the Cetera sequence. Given that difficult-to-clone regions are often in gene-rich areas of the human genome, isolation of the segments of the genome which are presently absent from the draft sequence becomes a biologically relevant task, and not simply an exercise in completing this large-scale project.

REFERENCES

Aravind L., Walker, D. R., Koonin, R. V., *Nucleic Acids. Res.* 27, 1223, 1999.

Battey, J., Jordan, E., Cox W. Dove., *Nature Genet.* 21, 73, 1999.

Craig, J. M., Bickmore, W. A., *Nature Genet* 23, 387, 1999.

Deloukas, P., *et al.*, *Science* 282, 744, 1998.

Edwards, J.H., *Ann. Hum. Genet.* 29, 77, 1965.

Eisen, J. A., Hanawalt, P.C., *Mutat. Res. DNA Repair* 459, 243, 2000.

Hofmann, K., *Cell. Mol. Life Sci.* 55, 1113, 1999.

International Human Genome Sequencing Consortium, *Nature* 409, 860, 2001.

Jackson, D. A., Iborra, F. J., Manders, E. M., Cook, P.R., *Mol. Biol.* 16, 54,1999.

Jones, N. L., CRS Long Report for Congress, RL30006, *Genetic Information: Legal Isuues Relating to Discrimination and Privacy*, 2 October 2000.

Marra M. A., *et al.*, *Genome Res.* 7, 1072, 1997.

Nahon, J. L., *Crit. Rev. Neurobiol.* 8, 221, 1994.

National Center for Genome Resources Survey, *Attitudes Toward Genetic Testing*, March 4, 1998.

Pandey, A. and Mann, M., *Nature* 405, 837, 2000.

Patthy, L., *Gene* 238, 103, 1999.

Peltonen, L., McKisick, V.A., *Science* 291, 1224, 2001.

Pavy, *N., et al.*, *Bioinformatics* 15, 887, 1999.

Sakahira, H., Enari, M., Nagat, S., *Nature*, 391, 96, 1998.

Schuler, G. D., *Trends Biotechnol*. 16, 456, 1998.

Venter J. C., *et al.*, *Science* 291, 1304, 2001.

Wansink, D. G., *et al.*, *J. Cell Biol*. 122, 283, 1993.

Watson J. D., and Crick, F.H.C., *Nature*, 171, 737, 1953.

Williams, N.M., *et al.*, *Hum. Mol. Genet*. 8, 1729, 1999.

7

Human Genome

A project with the goal of determining the complete nucleotide sequence of the human genome was first formally proposed in 1985. In subsequent years, the idea met with mixed reactions in the scientific community. In 1990, the Human Genome Project (HGP) was officially initiated in US under the direction of the National institutes of Health and US Department of energy. In 1998, HGP announced its intention to build a unique genome-sequencing facility, to determine the sequence of the human genome over a three-year period. The sequencing was performed by a whole-genome random shotgun method with subsequent assembly of the sequenced segments.

In 1977, when Sanger reported his method for determining the order of nucleotides of DNA using chain-terminating nucleotide analogs, the modern history of DNA sequencing began. In the same year, the first human gene was isolated and sequenced. In 1986, Hood and co-workers described an improvement in the Sanger sequencing method that included attaching fluorescent dyes to the nucleotides, which permitted them to be sequentially read by a computer. The first automated DNA sequencer, developed by applied Biosystems in California in 1987, was shown to be successful when the sequences of two genes were obtained with this new technology.

From early sequencing of human genomic regions, it became clear that cDNA sequences which are reverse-transcribed from RNA, would be essential to annotate and validate gene predictions in the human genome. These studies were the basis in part for the development of the expressed sequence tag (EST) method of gene identification, which is a random selection, very high throughput sequencing approach to characterise cDNA libraries. The EST method led to the rapid discovery and mapping of human genes. The increasing numbers of human EST sequences necessitated the development of new computer algorithms to analyse large amounts of sequence data, and in 1993 at The Institute for Genomic Research (TIGR), an algorithm was developed that permitted assembly and analysis of hundreds of thousands of ESTs. In 1994, when a microbial genome-sequencing project was contemplated at TIGR, a whole-genome shotgun sequencing approach was considered possible with the TIGR-EST assembly algorithm.

In 1995, the 1.8-Mbp *Haemophilus influenzae* genome was completed by a whole-genome shotgun sequencing method. The experience with several subsequent genome-sequencing efforts established the broad applicability of this approach. A key feature of the sequencing approach used for these megabase-size and larger genomes was the use of paired-end sequences (mate pairs), derived from subclone libraries with distinct insert sizes and cloning characteristics. Paired-end sequences are sequences 500 to 600 bp in length from both ends of double-stranded DNA clones of prescribed lengths.

The success of using end sequences from long segments (18 to 20 kbp) of DNA cloned into bacteriophage lambda in assembly of the microbial genomes led to the suggestion of an approach to simultaneously map and sequence the human genome by means of end sequences from 150-kbp bacterial artificial chromosomes (BACs). The end sequences spanned by known distances provide long-range continuity across the genome. A modification of the BAC end-sequencing (BES) method was applied successfully to complete chromosome 2 from the *arabidopsis thaliana* genome.

Weber and Myers proposed whole-genome shotgun sequencing of the human genome, in 1997. Their proposal was not well received. However, by early 1998, as less than five per cent of the genome had been sequenced. The rate of progress in human genome sequencing worldwide was very slow, and the prospects for finishing the genome by the 2005 goal were uncertain. In early 1998, Pe Biosystems developed an automated, high-throughput capillary DNA sequencer, subsequently called the ABI PRISM 3700 DNA analyser.

Discussions between Pe Biosystems and TIGR scientists resulted in a plan to undertake the sequencing of the human genome with the 3700 DNA analyser and the whole-genome shotgun sequencing techniques developed at TIGR. Many of the principles of operation of a genome-sequencing facility were established in the TiGR facility. However, the facility envisioned for Celera would have a capacity roughly 50 times that of TiGR, and thus new developments were required for sample preparation and tracking and for whole genome assembly. Some argued that the required 150-fold scale-up from the *H. influenzae* genome to the human genome with its complex repeat sequences was not feasible. The *Drosophila melanogaster* genome was thus chosen as a test case for whole-genome assembly on a large and complex eukaryotic genome.

In collaboration with Gerald Rubin and the Berkeley *Drosophila* Genome Project, the nucleotide sequence of the 120-Mbp euchromatic portion of the *Drosophila* genome was determined over a one-year period.

The *Drosophila* genome-sequencing effort resulted in two key findings:

i) The assembly algorithms could generate chromosome assemblies with highly accurate order and orientation with substantially less than 10-fold coverage, and

ii) undertaking multiple interim assemblies in place of one comprehensive final assembly was not of value.

These findings led to a modified whole-genome shotgun sequencing approach to the human genome.

The HGP initially proposed to do 10-fold sequence coverage of the genome over a three-year period and to make interim, assembled sequence data available quarterly. The modifications included a plan to perform random shotgun sequencing to ~5-fold coverage and to use the unordered and unoriented BAC sequence fragments and subassemblies published in GenBank by the publicly funded genome effort to accelerate the project.

Human genome sequencing was initiated on 8 September 1999 and completed on 17 June 2000. The first assembly was completed 25 June 2000, and the assembly reported here was completed on 1 October 2000. HGP developed two different assembly approaches for assembling the ~3 billion bp that make up the 23 pairs of chromosomes of the *Homo sapiens* genome. Any GenBank-derived data were shredded to remove potential bias to the final sequence from chimeric clones, foreign DNA contamination, or misassembled contigs.

DNA SEQUENCING

The plasmid library construction is the first critical step in shotgun sequencing. If the DNA libraries are not uniform in size non-chimeric, and do not randomly represent the genome, then the subsequent steps cannot accurately reconstruct the genome sequence. HGP used automated high-throughput DNA sequencing and the computational infrastructure to enable efficient tracking of enormous amounts of sequence information. Sequencing and tracking from both ends of plasmid clones from 2-, 10-, and 50-kbp libraries were essential to the computational reconstruction of the genome.

The Declaration of Helsinki offers recommendations for conducting experiments with human subjects. HGP convened an Institutional Review Board (IRB) that helped HGP to establish the protocol for obtaining and using human DNA and the informed consent process used to enroll research volunteers for the DNA-sequencing studies. HGP adopted several steps and procedures to protect the privacy rights and confidentiality of the research subjects. These included a two-stage consent process, a secure random alphanumeric coding system for specimens and records, circumscribed contact with the subjects by researchers, and options for off-site contact of donors.

In addition, Celera applied for and received a Certificate of Confidentiality from the Department of Health and Human Services. This Certificate authorised Celera to protect the privacy of the individuals who volunteered to be donors as provided in Section 301(d) of the Public Health Service Act 42 USC 241(d). Celera and the IRB believed that the initial version of a completed human genome should be a composite derived from multiple donors of diverse ethnic backgrounds. Prospective donors were asked, on a voluntary basis to self-designate an ethnogeographic category.

There are three fundamental items of information, from each donor were recorded and linked by confidential code to the donated sample such as age, sex, and self-designated ethnogeographic group. From females, ~130 ml of whole, heparinised blood was collected. From males, ~130 ml of whole, heparinised blood was collected, as well as five specimens of semen, collected over a six-week period. Permanent lymphoblastoid cell lines were created by epstein-Barr virus immortalisation. DNA from five subjects was selected for genomic DNA sequencing: two males and three females-one African-American, one Asian-Chinese, one Hispanic-Mexican, and two Caucasians. The decision of whose DNA to sequence was based on a complex mix of factors, including the goal of achieving diversity as well as technical issues like quality of the DNA libraries and availability of immortalised cell lines.

Library Construction

The preparation of high-quality plasmid libraries in a variety of insert sizes standing central to the whole-genome shotgun sequencing process. High-quality libraries have an equal representation of all parts of the genome, a small number of clones without inserts, and no contamination from such sources as the mitochondrial genome and *Escherichia coli* genomic DNA. DNA from each donor was used to construct plasmid libraries in one or more of three size classes: 2 kbp, 10 kbp, and 50 kbp. In designing the DNA-sequencing process, HGP focused on developing a simple system that could be implemented in a robust and reproducible manner.

Today, the sequencing protocols are based on the dideoxy sequencing method, which typically yields only 500 to 750 bp of sequence per reaction. This limitation on read length has made monumental gains in throughput a prerequisite for the analysis of large eukaryotic genomes. HGP accomplished this at the Celera facility, which occupies about 30,000 square feet of laboratory space and produces sequence data continuously at a rate of 175,000 total reads per day.

The DNA-sequencing facility is supported by a high-performance computational facility. The process for DNA sequencing was modular by design and automated. Inter-module sample backlogs allowed four principal modules to operate independently:

i) library transformation, plating, and colony picking;
ii) DNA template preparation;
iii) dideoxy sequencing reaction set-up and purification; and
iv) sequence determination with the ABI PRISM 3700 DNA analyser.

Since the inputs and outputs of each module have been carefully matched and sample backlogs are continuously managed, sequencing has proceeded without a single day's interruption since the initiation of the *Drosophila* project in May 1999.

The ABI 3700 is a fully automated capillary array sequencer and as such can be operated with a minimal amount of hands-on time, currently estimated at about 15 minuts per day. The capillary system also facilitates correct associations of sequencing traces with samples through the elimination of manual sample loading and lane-tracking errors associated with slab gels. About 65 production staff were hired and trained, and were rotated on a regular basis through the four production modules.

A central laboratory information management system (LIMS) tracked all sample plates by unique barcode identifiers. The facility was supported by a quality control team that performed raw material and in-process testing and a quality assurance group with responsibilities including document control, validation, and auditing of the facility.

Trace processing pipeline

To process each sequence file, an automated trace-processing pipeline has been developed. After quality and vector trimming, the average trimmed sequence length was 543 bp, and the sequencing accuracy was exponentially distributed with a mean of 99.5 per cent and with less than one in 1, 000 reads being less than 98 per cent accurate. Each trimmed sequence was screened for matches to contaminants including sequences of vector alone, *E. coli* genomic DNA, and human mitochondrial DNA. A total of 713 reads matched *E. coli* genomic DNA and 2, 114 reads matched the human mitochondrial genome. The entire read for any sequence with a significant match to a contaminant was discarded.

Quality Control

As the size and repetitive nature of the genome to be sequenced increases, the importance of the base-pair level accuracy of the sequence data also increases. Each sequence read must be placed uniquely in the genome, and even a modest error rate can reduce the effectiveness of assembly. Procedural controls were established for maintaining the validity of sequence mate-pairs as sequencing reactions proceeded through the process, including strict rules built into the LIMS.

The accuracy of sequence data produced by the Celera process was validated in the course of the Drosophila genome project. By collecting data for the entire human genome in a single facility HGP was able to ensure uniform quality standards and the cost advantages associated with automation, an economy of scale and process consistency.

Genome Assembly Strategy

This section discusses the two approaches that HGP used to assemble the genome. One method involves the computational combination of all sequence reads with shredded data from GenBank to generate an independent, non-biased view of the genome. The second approach involves clustering all of the fragments to a region or chromosome on the basis of mapping information. The clustered data were then shredded and subjected to computational assembly.

Both approaches provided essentially the same reconstruction of assembled DNA sequence with proper order and orientation. The second method provided slightly greater sequence coverage and was the principal sequence used for the analysis phase. In addition, HGP documented the completeness and correctness of this assembly process and provide a comparison to the public genome sequence, which was reconstructed largely by an independent BAC-by-BAC approach.

HGP assemblies effectively covered the euchromatic regions of the human chromosomes. More than 90 per cent of the genome was in scaffold assemblies of 100,000 bp or greater, and 25 per cent of the genome was in scaffolds of 10 million bp or larger. Shotgun sequence assembly is a classic example of an inverse problem: given a set of reads randomly sampled from a target sequence, reconstruct the order and the position of those reads in the target.

Genome assembly algorithms developed for *Drosophila* have now been extended to assemble the ~25-fold larger human genome. Celera assemblies consist of a set of contigs that are ordered and oriented into scaffolds that are then mapped to chromosomal locations by using known markers. The contigs consist of a collection of overlapping sequence reads that provide a consensus reconstruction for a contiguous interval of the genome.

Mate-pairs are a central component of the assembly strategy. They are used to produce scaffolds in which the size of gaps between consecutive contigs is known with reasonable precision. This is accomplished by observing that a pair of reads, one of which is in one coring, and the other of which is in another, implies an orientation and distance between the two contigs.

The assemblies did not incorporate all reads into the final set of reported scaffolds. This set of unincorporated reads is termed 'chaff' and typically consisted of reads from within highly repetitive regions, data from other organisms introduced through various routes as found in many genome projects, and data of poor quality.

Data sets

HGP used two independent sets of data for genome assemblies.

The first was a random shotgun data set of 27.27 million reads of average length 543 bp produced at Celera. This consisted largely of mate-pair reads from 16 libraries constructed from DNA samples taken from five different donors. Libraries with insert sizes of 2, 10, and 50 kbp were used. By looking at how mate-pairs from a library were positioned in known sequenced stretches of the genome, HGP was able to characterise the range of insert sizes in each library and determine a mean and standard deviation.

The clone coverage is the coverage of the genome in cloned DNA, considering the entire insert of each clone that has sequence from both ends. The clone coverage provides a measure of the amount of physical DNA coverage of the genome. Assuming a genome size of 2.9 Gbp, the Celera trimmed sequences gave a 5.1X coverage of the genome, and clone coverage (for a total of 38.7X) was 3.42X, 16.40X, and 18.84X for the 2-, 10-, and 50-kbp libraries, respectively.

The second set of data was from the publicly funded Human Genome Project (PFP) and is primarily derived from BAC clones. The BAC data input to the assemblies came from a download of GenBank on 1 September 2000 totalling 4, 443.3 Mbp of sequence. The data for each BAC is deposited at one of four levels of completion.

i) *Phase 0 data:* These are a set of generally unassembled sequencing reads from a very light shotgun of the BAC, typically less than 1X.
ii) *Phase 1 data:* These are unordered assemblies of contigs, which HGP calls BAC contigs or bactigs.
iii) *Phase 2 data:* These are ordered assemblies of bactigs.
iv) *Phase 3 data:* These are complete BAC sequences.

In the past two years the PFP has focused on a product of lower quality and completeness, but on a faster time-course, by concentrating on the production of Phase 1 data from a 3X to 4X fight-shotgun of each BAC clone.

HGP screened the bactig sequences for contaminants by using the BLAST algorithm against three data sets:

i) vector sequences in univec core, filtered for a 25-bp match at 98 per cent sequence identity at the ends of the sequence and a 30-bp match internal to the sequence;
ii) the non-human portion of the High Throughput Genomic (HTG) Sequences division of GenBank, filtered at 200 bp at 98 per cent; and
iii) the non-redundant nucleotide sequences from GenBank without primate and human virus entries, filtered at 200 bp at 98 per cent.

Whenever 25 bp or more of vector was found within 50 bp of the end of a contig, the tip up to the matching vector was excised. Under these criteria HGP removed 2.6 Mbp of possible contaminant and vector from the Phase 3 data, 61.0 Mbp from the Phase 1 and 2 data, and 16.1 Mbp from the Phase 0 data. This left with a total of 4363.7 Mbp of PFP sequence data 20 per cent finished, 75 per cent rough-draft, and five per cent single sequencing reads. An additional 104,018 BAC end-sequence mate-pairs were also downloaded and included in the data sets for both assembly processes.

Strategies

HGP followed two different approaches to assembly strategies. The first was a whole-genome assembly process that used Celera data and the PFP data in the form of additional synthetic shotgun data, and the second was a compartmentalised assembly process that first partitioned the Celera and PFP data into sets localised to large chromosomal segments and then performed *ab initio* shotgun assembly on each set.

For the first approach, the whole-genome assembly, the PFP data was first disassembled or shredded into a synthetic shotgun data set of 550-bp reads that form a perfect 2X covering of the bactigs. This resulted in 16.05 million faux reads that were sufficient to cover the genome 2.96X because of redundancy in the BAC data set without incorporating the biases inherent in the PEP assembly process. The combined data set of 43.32 million reads (8X), and all associated mate-pair information, were then subjected to HGP's whole-genome assembly algorithm to produce a reconstruction of the genome.

Neither the location of a BAC in the genome nor its assembly of bactigs was used in this process. Bactigs were shredded into reads because HGP found strong evidence that 2.13 per cent of them were misassembled. Furthermore, BAC location information was ignored because some BACs were not correctly placed on the PFP physical map and because HGP found strong evidence that at least 2.2 per cent of the BACs contained sequence data that were not part of the given BAC, possibly as a result of sample-tracking errors.

In the second approach, compartmentalised shotgun assembly (CSA), Celera and PFP data were partitioned into the largest possible chromosomal segments or components that could be determined with confidence, and then shotgun assembly was applied to each partitioned subset wherein the bactig data were again shredded into faux reads to ensure an independent *ab initio* assembly of the component. By subsetting the data in this way, the overall computational effort was reduced and the effect of interchromosomal duplications was ameliorated. This also resulted in a reconstruction of the genome that was relatively independent of the whole-genome assembly results so that the two assemblies could be compared for consistency.

The quality of the partitioning into components was crucial so that different genome regions were not mixed together. HGP constructed components from:

i) the longest scaffolds of the sequence from each BAC, and
ii) assembled scaffolds of data unique to Celera's data set.

The BAC assemblies were obtained by a combining assembler that used the bactigs and the 5X Celera data mapped to those bactigs as input. This effort was undertaken as an interim step solely because the more accurate and complete the scaffold for a given sequence stretch, the more accurately one can tile these scaffolds into contiguous components on the basis of sequence overlap and mate-pair information. HGP further visually inspected and curated the scaffold tiling of the components to further increase its accuracy. For the final CSA assembly, all but the partitioning was ignored, and an independent, *ab initio* reconstruction of the sequence in each component was obtained by applying the whole-genome assembly algorithm to the partitioned, relevant Celera data and the shredded faux reads of the partitioned relevant bactig data.

Whole-genome assembly

Whole-genome assembler consists of a pipeline composed of five principal stages:

i) screener,
ii) overlapper,
iii) unitigger,
iv) scaffolder, and
v) repeat resolver.

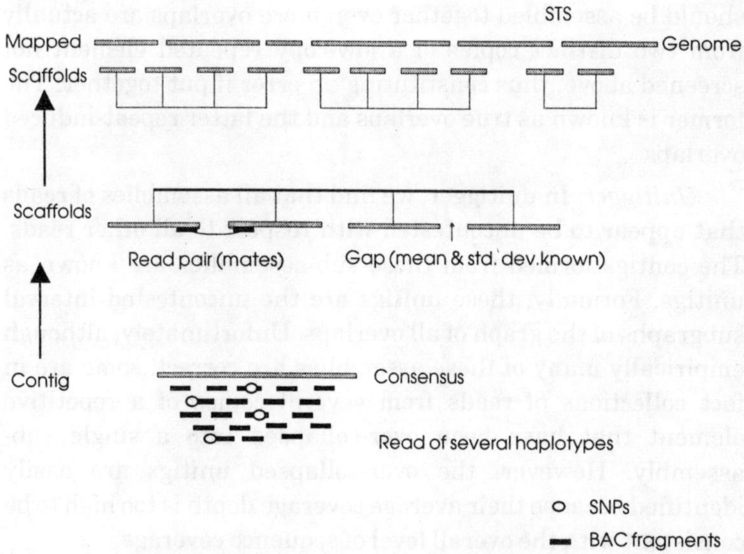

Figure 7. 1. Anatomy of whole-genome assembly.

Screener: The screener finds and marks all microsatellite repeats with less than a 6-bp element, and screens out all known interspersed repeat elements, including alu, Line, and ribosomal DNA. Marked regions get searched for overlaps, whereas screened regions do not get searched, but can be part of an overlap that involves unscreened matching segments.

Overlapper: The overlapper compares every read against every other read in search of complete end-to-end overlaps of at least 40 bp and with no more than six per cent differences in the match. Since all data are scrupulously vector-trimmed, the overlapper can insist on complete overlap matches. Computing the set of all overlaps took roughly 10, 000 CPU hours with a suite of four-processor alpha SMPs with five gigabytes of RAM. This took four to five days in elapsed time with 40 such machines operating in parallel. Every overlap computed above is statistically a 1-in-10^{17} event and thus not a coincidental event. What makes assembly combinatorially difficult is that while many overlaps are actually sampled from overlapping regions of the genome, and thus imply that the sequence reads should be assembled together even more overlaps are actually from two distinct copies of a low-copy repeated element not screened above, thus constituting an error if put together. The former is known as true overlaps and the latter repeat-induced overlaps.

Unitigger: In unitigger, we find that all assemblies of reads that appear to be uncontested with respect to all other reads. The contigs formed from these sub-assemblies are known as unitigs. Formally, these unitigs are the uncontested interval subgraphs of the graph of all overlaps. Unfortunately, although empirically many of these assemblies are correct, some are in fact collections of reads from several copies of a repetitive element that have been over-collapsed into a single sub-assembly. However, the over-collapsed unitigs are easily identified because their average coverage depth is too high to be consistent with the overall level of sequence coverage.

HGP developed a simple statistical discriminator that gives the logarithm of the odds ratio that a unitig is composed of unique DNA or of a repeat consisting of two or more copies. The discriminator, set to a sufficiently stringent threshold, identifies a subset of the unitigs. In addition a second, less stringent threshold identifies a subset of remaining unitigs very likely to be correctly assembled, of which we select those that will consistently scaffold, and thus are again almost certain to be correct. We have the union of these two sets u-unitigs. Empirically, from a 6X simulated shotgun of human chromosome 22 that they get the u-unitigs covering 98 per cent of the stretches of unique DNA that are >2 kbp long. We further able to identify the boundary of the start of a repetitive element at the ends of a u-unitig and leverage this so that u-unitigs span more than 93 per cent of all singly interspersed alu elements and other. Significance of running the unitigger was thus a set of correctly assembled subcontigs covering an estimated 73.6 per cent of the human genome.

Scaffolder: The scaffolder proceeded to use mate-pair information to link these together into scaffolds. When there are two or more mate pairs that imply that a given pair of u-unitigs are at a certain distance and orientation with respect to each other, the probability of this being wrong is again roughly one in 10^{10}, assuming that mate pairs are false less than 2 per cent of the time. Thus, one can with high confidence link together all u-unitigs that are linked by at least two two or 10-kbp mate pairs producing intermediate-sized scaffolds that are then recursively linked together by confirming 50-kbp mate pairs and BAC end sequences. This process yielded scaffolds that are on the order of megabase pairs in size with gaps between their contigs that generally correspond to repetitive elements and occasionally to small sequencing gaps. These scaffolds reconstruct the majority of the unique sequence within a genome.

Repeat resolver: For *Drosophila* assembly, HGP was engaged in a three-stage repeat resolution strategy, where each stage was progressively more aggressive and thus more likely to make a mistake. For the human assembly, HGP continued to use the first 'rocks' substage where all unitigs with a good, but not definitive, discriminator score are placed in a scaffold gap. This was done with the condition that two or more mate pairs with one of their reads already in the scaffold unambiguously place the unitig in the given gap. HGP estimated the probability of inserting a unitig into an incorrect gap with this strategy to be less than 10^{-7} based on a probabilistic analysis.

HGP revised the ensuing 'stones' substage of the human assembly, making it more like the mechanism suggested in an earlier work. For each gap, every read R that is placed in the gap by virtue of its mated pair M being in a contig of the scaffold and implying R's placement is collected. Celera's mate-pairing information is correct more than 99 per cent of the time. Thus almost every, but not all, of the reads in the set belong in the gap, and when a read does not belong it rarely agrees with the remainder of the reads. Therefore, we simply assemble this set of reads within the gap, eliminating any reads that conflict with the assembly. This operation proved much more reliable than the one it replaced for the *Drosophila* assembly; in the assembly of a simulated shotgun data set of human chromosome 22, all stones were placed correctly.

HGP used a final method of resolving gaps—to fill them with assembled BAC data that cover the gap. This external gap is called as 'walking'. HGP did not include the very aggressive 'pebbles' substage described in *Drosophila* work, which made enough mistakes so as to produce repeat reconstructions for long interspersed elements whose quality was only 99.62 per cent correct. HGP decided that for the human genome it was philosophically better not to introduce a step that was certain to produce less than 99.99 per cent accuracy.

A consensus sequence of every contig is produced at the final stage of the assembly process, and also at several intermediate points. HGP's algorithm is driven by the principle of maximum parsimony, with quality value-weighted measures for evaluating each base. The net effect is a Bayesian estimate of the correct base to report at each position. Consensus generation uses Celera data whenever it is present.

One key element in achieving a WGA of the human genome was to parallelise the overlapper and the central consensus sequence-constructing subroutines. In addition, memory was a real issue—a straightforward application of the software HGP had built for *Drosophila* would have required a computer with a 600-gigabyte RAM. By making the overlapper and unitigger incremental, HGP was able to achieve the same computation with a maximum of instantaneous usage of 28 gigabytes of RAM.

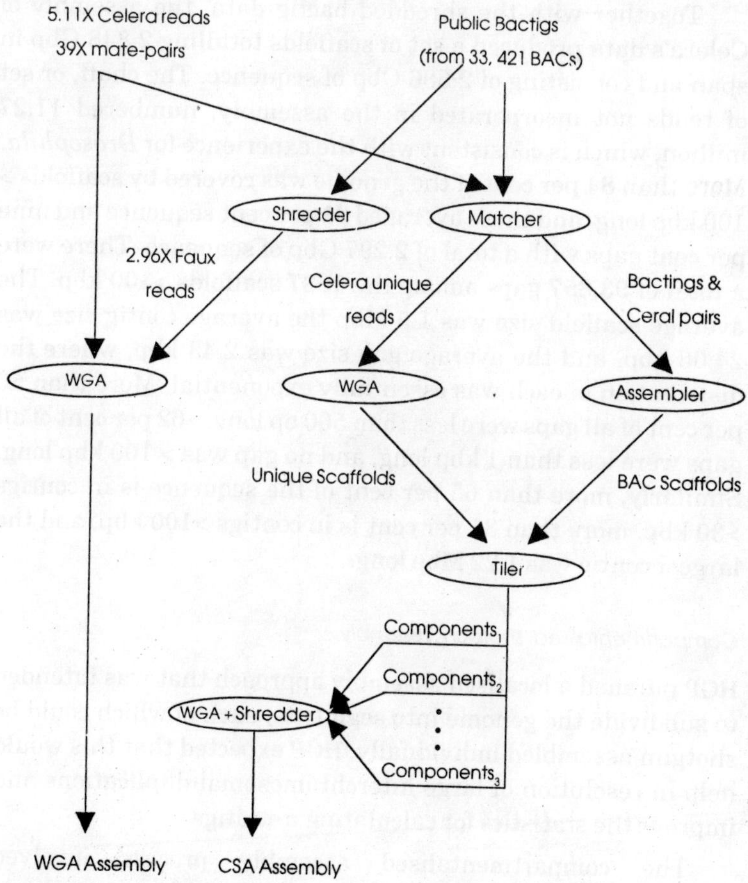

Figure 7. 2. Architecture of Celera's two-pronged assembly strategy.

Moreover, the incremental nature of the first three stages allowed HGP to continually update the state of this part of the computation as data were delivered and then perform a seven-day run to complete Scaffolding and Repeat Resolution whenever desired. For HGP's assembly operations, the total compute infrastructure consists of 10 four-processor SMPs with four gigabytes of memory per cluster and a 16-processor NUMA machine with 64 gigabytes of memory.

Together with the shredded bactig data, the assembly of Celera's data produced a set of scaffolds totalling 2.848 Gbp in span and consisting of 2.586 Gbp of sequence. The chaff, or set of reads not incorporated in the assembly, numbered 11.27 million, which is consistent with the experience for *Drosophila*. More than 84 per cent of the genome was covered by scaffolds > 100 kbp long, and these averaged 91 per cent sequence and nine per cent gaps with a total of 2.297 Gbp of sequence. There were a total of 93,857 gaps among the 1637 scaffolds >100 kbp. The average scaffold size was 1.5 Mbp the average contig size was 24.06 kbp, and the average gap size was 2.43 kbp, where the distribution of each was essentially exponential. More than 50 per cent of all gaps were less than 500 bp long, >62 per cent of all gaps were less than 1 kbp long, and no gap was > 100 kbp long. Similarly, more than 65 per cent of the sequence is in contigs >30 kbp, more than 31 per cent is in contigs >100 kbp, and the largest contig was 1.22 Mbp long.

Compartmentalised shotgun assembly

HGP pursued a localised assembly approach that was intended to subdivide the genome into segments, each of which could be shotgun assembled individually. HGP expected that this would help in resolution of large interchromosomal duplications and improve the statistics for calculating u-unitigs.

The compartmentalised assembly process involved clustering Celera reads and bactigs into large, multiple megabase regions of the genome, and then running the WGA assembler on the Celera data and shredded, faux reads obtained from the bactig data. The first phase of the CSA strategy was to separate Celera reads into those that matched the BAC contigs for a particular PFP BAC entry, and those that did not match any public data. Such matches must be guaranteed to properly place a Celera read, so all reads were first masked against a library of common repetitive elements, and only matches of at least 40 bp to unmasked portions of the read constituted a hit.

Of Celera's 27.27 million reads, 20.76 million matched a bactig and another 0.62 million reads, which did not have any matches, were nonetheless identified as belonging in the region of the bactig's BAC because their mate matched the bactig. Of the remaining reads, 2.92 million were completely screened out and so could not be matched, but the other 2.97 million reads had unmasked sequence totalling 1. 189 Gbp that were not found in the GenBank data set. Since the Cetera data are 5.11X redundant, HGP estimated that 240 Mbp of unique Celera sequence is not in the GenBank data set.

In the second phase of the CSA process, a combining assembler took the relevant 5X Cetera reads and bactigs for a BAC entry, and produced an assembly of the combined data for that locale. These high-quality sequence reconstructions were a transient result whose utility was simply to provide more reliable information for the purposes of their tiling into sets of overlapping and adjacent scaffold sequences in the next step.

In outline, the combining assembler first examines the set of matching Celera reads to determine if there are excessive pileups indicative of unscreened repetitive elements. Wherever these occur, reads in the repeat region whose mates have not been mapped to consistent positions are removed. Then all sets of mate pairs that consistently imply the same relative position of two bactigs are bundled into a link and weighted according to the number of mates in the bundle.

A 'greedy' strategy then attempts to order the bactigs by selecting bundles of mate-pairs in order of their weight. A selected mate-pair bundle can tie together two formative scaffolds. It is incorporated to form a single scaffold only if it is consistent with the majority of links between contigs of the scaffold. Once scaffolding is complete, gaps are filled by the 'stones' strategy for the WGA assembler. The GenBank data for the Phase 1 and 2 BACs consisted of an average of 19.8 bactigs per BAC of average size 8, 099 bp.

Application of the combining assembler resulted in individual Celera BAC assemblies being put together into an average of 1.83 scaffolds consisting of an average of 8.57 contigs of average size 18,973 bp. In addition to defining order and orientation of the sequence fragments, there were 57 per cent fewer gaps in the combined result.

For phase 0 data, the average GenBank entry consisted of 91.52 reads of average length 784 bp. Application of the combining assembler resulted in an average of 54.8 scaffolds consisting of an average of 58.1 contigs of average size 873 bp. Basically, some small amount of assembly took place, but not enough Celera data were matched to truly assemble the 0.5X to 1X data set represented by the typical Phase 0 BACs. The combining assembler was also applied to the Phase 3 BACs for SNP identification, confirmation of assembly, and localisation of the Celera reads.

The phase 0 data suggest that a combined whole-genome shotgun data set and 1X light-shotgun of BACs will not yield good assembly of BAC regions; at least 3X light-shotgun of each BAC is needed. The 5.89 million Celera fragments not matching the GenBank data were assembled with whole-genome assembler. The assembly resulted in a set of scaffolds totaling 442 Mbp in span and consisting of 326 Mbp of sequence. More than 20 per cent of the scaffolds were >5 kbp long, and these averaged 63 per cent sequence and 27 per cent gaps with a total of 302 Mbp of sequence. All scaffolds >5 kbp were forwarded along with all scaffolds produced by the combining assembler to the subsequent tiling phase. At this stage, HGP typically had one or two scaffolds for every BAC region constituting at least 95 per cent of the relevant sequence, and a collection of disjoint Celera-unique scaffolds.

The next step in developing the genome components was to determine the order and overlap tiling of these BAC and Celera-unique scaffolds across the genome. For this, HGP used Celera's 50-kbp mate-pairs information, and BAC-end pairs and sequence tagged site (STS) markers to provide long-range guidance and chromosome separation. Given the relatively manageable number of scaffolds, HGP chose not to produce this tiling in a fully automated manner, but to compute an initial tiling with a good heuristic and then use human curators to resolve discrepancies or missed join opportunities.

To this end, HGP developed a graphical user interface that displayed the graph of tiling overlaps and the evidence for each. A human curator could then explore the implication of mapped STS data dot-plots of sequence overlap, and a visual display of the mate-pair evidence supporting a given choice. The result of this process was a collection of 'components', where each component was a tiled set of BAC and Celera-unique scaffolds that had been curator-approved.

HGP assembled each component with the WGA algorithm, in order to generate the final CSA. As was done in the WGA process, the bactig data were shredded into a synthetic 2X shotgun data set in order to give the assembler the freedom to independently assemble the data.

By using faux reads rather than bactigs, the assembly algorithm could correct errors in the assembly of bactigs and remove chimeric content in a PFP data entry. Chimeric or contaminating sequence would not be incorporated into the reassembly of the component because it did not belong there.

In effect, the previous steps in the CSA process served only to bring together Celera fragments and PFP data relevant to a large contiguous segment of the genome, wherein HGP applied the assembler used for WGA to produce an *ab initio* assembly of the region. WGA assembly of the components resulted in a set of scaffolds totaling 2.906 Gbp in span and consisting of 2.654 Gbp of sequence. The chaff, or set of reads not incorporated into the assembly, numbered 6.17 million, or 22 per cent. More than 90.0 per cent of the genome was covered by scaffolds spanning >100 kbp long, and these averaged 92.2 per cent sequence and 7.8 per cent gaps with a total of 2.492 Gbp of sequence.

There were a total of 105,264 gaps among the 107,199 contigs that belong to the 1940 scaffolds spanning >i 00 kbp.

The average scaffold size was 1.4 Mbp, the average contig size was 23.24 kbp, and the average gap size was 2.0 kbp where each distribution of sizes was exponential. As such, averages tend to be under-representative of the majority of the data. Consider also that more than 49 per cent of all gaps were <500 bp long, more than 62 per cent of all gaps were <1 kbp, and all gaps are < 100 kbp long. Similarly, more than 73 per cent of the sequence is in contigs > 30 kbp, more than 49 per cent is in contigs >i 00 kbp, and the largest contig was 1.99 Mbp long.

HGP compared scaffolds from WGA and CSA assemblies as another means of investigating their completeness, consistency, and contiguity. From each assembly, a set of reference scaffolds containing at least 1, 000 fragments was obtained. This amounted to 2, 218 WGA scaffolds and 1, 717 CSA scaffolds, for a total of 2.087 Gbp and 2.474 Gbp. The sequence of each reference scaffold was compared to the sequence of all scaffolds from the other assembly with which it shared at least 20 fragments or at least 20 per cent of the fragments of the smaller scaffold.

For each such comparison, all matches of at least 200 bp with at most two per cent mismatch were tabulated. From this tabulation, HGP estimated the amount of unique sequence in each assembly in two ways. The first was to determine the number of bases of each assembly that were not covered by a matching segment in the other assembly. Some 82.5 Mbp of the WGA was not covered by the CSA, whereas 204.5 Mbp of the CSA was not covered by the WGA. This estimate did not require any consistency of the assemblies or any uniqueness of the matching segments.

Thus another analysis was conducted in which matches of less than one kbp between a pair of scaffolds were excluded unless they were confirmed by other matches having a consistent order and orientation. This gives some measure of consistent coverage: 1.982 Gbp of the WGA is covered by the CSA and 2.169 Gbp of the CSA is covered by the WGA by this more stringent measure. The comparison of WGA to CSA also permitted evaluation of scaffolds for structural inconsistencies.

HGP looked for instances in which a large section of a scaffold from one assembly matched only one scaffold from the other assembly, but failed to match over the full length of the overlap implied by the matching segments. An initial set of candidates was identified automatically, and then each candidate was inspected by hand.

From this process, HGP identified 31 instances in which the assemblies appear to disagree in a non-local fashion. These cases are being further evaluated to determine which assembly is in error and why. In addition, HGP evaluated local inconsistencies of order or orientation. The following results exclude cases in which one contig in one assembly corresponds to more than one overlapping contig in the other assembly. Most of these small rearrangements involved segments on the order of hundreds of base pairs and rarely >1 kbp. HGP found a total of 295 kbp in the CSA assemblies that were locally inconsistent with the WGA assemblies, whereas 2.108 Mbp in the WGA assembly were inconsistent with the CSA assembly.

In terms of coverage the CSA assembly was slightly more consistent than the WGA, because it was in effect performing a few thousand shotgun assemblies of megabase-sized problems, whereas the WGA is performing a shotgun assembly of a gigabase-sized problem. When one considers the increase of two-and-a-half orders of magnitude in problem size, the information loss between the two is remarkably small.

Mapping Scaffolds

To order and orient the scaffolds on the chromosomes was the final step in assembling the genome. HGP firstly grouped scaffolds together on the basis of their order in the components from CSA. These grouped scaffolds were reordered by examining residual mate-pairing data between the scaffolds. HGP next mapped the scaffold groups onto the chromosome using physical mapping data. This step depends on having reliable high-resolution map information such that each scaffold will overlap multiple markers.

There are two genome-wide types of map information available: high-density STS maps and fingerprint maps of BAC clones developed at Washington university. Among the genome-wide STS maps, GeneMap99 (GM99) has the most markers and therefore was most useful for mapping scaffolds.

The two different mapping approaches are complementary to one another. The fingerprint maps should have better local order because they were built by comparison of overlapping BAC clones. On the other hand, GM99 should have a more reliable long-range order, because the framework markers were derived from well-validated genetic maps. Both types of maps were used as a reference for human curation of the components that were the input to the regional assembly.

To determine the effectiveness of the fingerprint maps and GM99 for mapping scaffolds, HGP examined the reliability of these maps by comparison with large scaffolds. Only one per cent of the STS markers on the 10 largest scaffolds were mapped on a different chromosome on GM99. Two per cent of the STS markers disagreed in position by more than five framework bins. However, for the fingerprint maps, a two per cent chromosome discrepancy was observed, and on average 23.8 per cent of BAC locations in the scaffold sequence disagreed with fingerprint map placement by more than five BACs.

When further examining the source of discrepancy, it was found that most of the discrepancy came from four of the 10 scaffolds, indicating this there is variation in the quality of either the map or the scaffolds. All four scaffolds were assembled, as well as the other six, as judged by clone coverage analysis, and showed the same low discrepancy rate to GM99, and thus HGP concluded that the fingerprint map global order in these cases was not reliable. Smaller scaffolds had a higher discordance rate with GM99, but a lower discordance rate with the fingerprint maps.

This observation agrees with the clone coverage analysis that Celera scaffold construction was better supported by long-range mate pairs in larger scaffolds than in small scaffolds. HGP created two orderings of Celera scaffolds on the basis of the markers (BAC or STS) on these maps. Where the order of scaffolds agreed between GM99 and the Washu BAC map, HGP had a high degree of confidence that order was correct; these scaffolds were termed 'anchor scaffolds'. Only scaffolds with a low overall discrepancy rate with both maps were considered anchor scaffolds. Scaffolds in GM99 bins were allowed to permute in their order to match Washu ordering, provided they did not violate their framework orders.

Orientation of individual scaffolds was determined by the presence of multiple mapped markers with consistent order. Scaffolds with only one marker have insufficient information to assign orientation. HGP found 70.1 per cent of the genome in anchored scaffolds, more than 99 per cent of which are also oriented. Since GM99 is of lower resolution than the Washu map, a number of scaffolds without STS matches could be ordered relative to the anchored scaffolds because they included sequence from the same or adjacent BACs on the Washu map.

On the other hand, because of occasional Washu global ordering discrepancies, a number of scaffolds determined to be unmappable on the Washu map could be ordered relative to the anchored scaffolds with GM99. These scaffolds were termed 'ordered scaffolds'. HGP found that 13.9 per cent of the assembly could be ordered by these additional methods.

All scaffolds that could be placed between anchors were assigned to the interval between the anchored scaffolds and were deemed to be 'bounded' between them. For example, small scaffolds having STS hits from the same GeneMap bin or hitting the same BAC cannot be ordered relative to each other, but can be assigned a placement boundary relative to other anchored or ordered scaffolds. The remaining scaffolds either had no localisation information, conflicting information, or could only be assigned to a generic chromosome location.

HGP assigned a location for each scaffold placed on the chromosome by spreading out the scaffolds per chromosome.

The remaining unmapped scaffolds, constituting 2 per cent of the genome, were distributed evenly across the genome. By dividing the sum of unmapped scaffold lengths with the sum of the number of mapped scaffolds, HGP arrived at an estimate of interscaffold gap of 1483 bp. This gap was used to separate all the scaffolds on each chromosome and to assign an offset in the chromosome.

HGP encountered many problems that resulted in additional quality assessment and validation analysis, during the scaffold-mapping effort. At least 978 BACs were believed to have sequence data from more than one location in the genome. These BACs could not be assigned to unique positions within the CSA assembly and thus could not be used for ordering scaffolds.

Likewise, it was not always possible to assign STSs to unique locations in the assembly because of genome duplications, repetitive elements, and pseudogenes. Due to the time required for an exhaustive search for a perfect overlap, CSA generated 21.607 intrascaffold gaps where the mate-pair data suggested that the contigs should overlap, but no overlap was found.

These gaps were defined as a fixed 50 bp in length and make up 18.6 per cent of the total 116,442 gaps in the CSA assembly. HGP chose not to use the order of exons implied in cDNA or EST data as a way of ordering scaffolds. The rationale for not using this data was that doing so would have biased certain regions of the assembly by rearranging scaffolds to fit the transcript data and made validation of both the assembly and gene definition processes more difficult.

Validation Analysis

HGP analysed the assembly of the genome from the perspectives of completeness and correctness.

Completeness is the percentage of the euchromatic sequence represented in the assembly. This cannot be known with absolute certainty until the euchromatin sequence has been completed. However, it is possible to estimate completeness on the basis of:

i) the estimated sizes of intrascaffold gaps;

ii) coverage of the two published chromosomes, 21 and 22; and

iii) analysis of the percentage of an independent set of random sequences (STS markers) contained in the assembly.

The whole-genome libraries contain heterochromatic sequence and, although no attempt has been made to assemble it, there may be instances of unique sequence embedded in regions of heterochromatin as were observed in *Drosophila.* The sequences of human chromosomes 21 and 22 have been completed to high quality and published.

Although this sequence served as input to the assembler, the finished sequence was shredded into a shotgun data set so that the assembler had the opportunity to assemble it differently from the original sequence in the case of structural polymorphisms or assembly errors in the BAC data. In particular, the assembler must be able to resolve repetitive elements at the scale of components, and so this comparison reveals the level to which the assembler resolves repeats.

In certain areas, the assembly structure differs from the published versions of chromosomes 21 and 22. The consequence of the flexibility to assemble 'finished' sequence differently on the basis of Celera data resulted in an assembly with more segments than the chromosome 21 and 22 sequences. HGP examined the reasons why there are more gaps in the Celera sequence than in chromosomes 21 and 22 and expect that they may be typical of gaps in other regions of the genome.

In the Celera assembly, there are 25 scaffolds, each containing at least 10 kb of sequence, that collectively span 94.3 per cent of chromosome 21. Sixty-two scaffolds span 95.7 per cent of chromosome 22. The total length of the gaps remaining in the Celera assembly for these two chromosomes is 3.4 Mbp. These gap sequences were analysed by RepeatMasker and by searching against the entire genome assembly. About 50 per cent of the gap sequence consisted of common repetitive elements identified by RepeatMasker; more than half of the remainder was lower copy number repeat elements.

Another way of assessing completeness is to measure the content of an independent set of sequence data in the assembly. HGP compared 48,938 STS markers from Genemap99 to the scaffolds. Since these markers were not used in the assembly processes, they provided a truly independent measure of completeness EPCR and BLAST were used to locate STSs on the assembled genome.

HGP found 44, 524 (91 per cent) of the STSs in the mapped genome. An additional 2, 648 markers (5.4 per cent) were found by searching the unassembled data or 'chaff'. HGP identified 1, 283 STS markers (2.6 per cent) not found in either Celera sequence or BAC data as of September 2000, raising the possibility that these markers may not be of human origin.

If that were the case, the Celera assembled sequence would represent 93.4 per cent of the human genome and the unassembled data 5.5 per cent, for a total of 98.9 per cent coverage. Similarly, HGP compared CSA against 36,678 TNG radiation hybrid markers using the same method. HGP found that 32, 371 markers (88 per cent) were located in the mapped CSA scaffolds, with 2, 055 markers (5.6 per cent) found in the remainder. This gave a 94 per cent coverage of the genome through another genome-wide survey.

Correctness is defined as the structural and sequence accuracy of the assembly. Since the source sequences for the Celera data and the GenBank data are from different individuals, HGP could not directly compare the consensus sequence of the assembly against other finished sequence for determining sequencing accuracy at the nucleotide level, although this has been done for identifying polymorphisms.

The accuracy of the consensus sequence is at least 99.96 per cent on the basis of a statistical estimate derived from the quality values of the underlying reads. The structural consistency of the assembly can be measured by mate-pair analysis. In a correct assembly, every mated pair of sequencing reads should be located on the consensus sequence with the correct separation and orientation between the pairs. A pair is termed valid when the reads are in the correct orientation and the distance between them is within the mean ± 3 standard deviations of the distribution of insert sizes of the library from which the pair was sampled.

A pair is termed 'mis-oriented' when the reads are not correctly oriented, and is termed 'mis-separated' when the distance between the reads is not in the correct range but the reads are correctly oriented. The mean ± the standard deviation of each library used by the assembler was determined as described above. To validate these, HGP examined all reads mapped to the finished sequence of chromosome 21 and determined how many incorrect mate pairs there were as a result of laboratory tracking errors and chimerism and how tight the distribution of insert sizes was for those that were correct.

The standard deviations for all Celera libraries were quite small, less than 15 per cent of the insert length, with the exception of a few 50-kbp libraries. The 2- and 10-kbp libraries contained less than two per cent invalid mate pairs, whereas the 50-kbp libraries were somewhat higher. Thus, although the mate-pair information was not perfect, its accuracy was such that measuring valid, misoriented, and mis-separated pairs with respect to a given assembly was deemed to be a reliable instrument for validation purposes, especially when several mate pairs confirm or deny an ordering.

The clone coverage of the genome was 39X, meaning that any given base pair was, on average, contained in 39 clones or, equivalently, spanned by 39 mate-paired reads. Areas of low clone coverage or areas with a high proportion of invalid mate pairs would indicate potential assembly problems. HGP computed the coverage of each base in the assembly by valid mate pairs. For scaffolds >30 kbp in length, less than one per cent of the Cetera assembly was in regions of less than 3X clone coverage. Thus, more than 99 per cent of the assembly, including order and orientation, is strongly supported by this measure alone.

HGP examined the locations and number of all misoriented and mis-separated mates. In addition to doing this analysis on the CSA assembly, a study of the PFP assembly as of 5 September 2000 also performed. In this latter case, Celera mate pairs had to be mapped to the PFP assembly.

To avoid mapping errors due to high-fidelity repeats, the only pairs mapped were those for which both reads matched at only one location with less than six per cent differences. A threshold was set such that sets of five or more simultaneously invalid mate pairs indicated a potential breakpoint, where the construction of the two assemblies differed.

The graphic comparison of the CSA chromosome 21 assembly with the published sequence serves as a validation of this methodology. Blue tick-marks in the panels indicate breakpoints. There were a similar number of breakpoints on both chromosome sequences.

The exception was 12 sets of scaffolds in the Celera assembly that were mapped to the wrong positions because they were too small to be mapped reliably. There was a higher percentage of mis-oriented and mis-separated mate pairs in the large-insert libraries than in the small-insert libraries in both assemblies. The large-insert libraries are more likely to identify discrepancies simply because they span a larger segment of the genome.

There are many more breakpoints for the PFP assembly than for the Celera assembly. The order and orientation of Celera's assembly shows substantially fewer breakpoints except on the two finished chromosomes. In the CSA assembly, the size of all gaps have been estimated on the basis of the mate-pair data. Breakpoints can be caused by structural polymorphisms, because the two assemblies were derived from different human genomes.

GENE PREDICTION

HGP developed an integrated, evidence-based approach named Otto to enumerate the gene inventory. The evidence used to increase the likelihood of identifying genes includes regions conserved between the mouse and human genomes, similarity to ESTs or other m-RNA-derived data, or similarity to other proteins. A comparison of Otto (combined Otto-RefSeq and Otto-homology) with Genscan, a standard gene-prediction algorithm, showed greater sensitivity and specificity of Otto in the ability to define gene structure. Otto-predicted genes were complemented with a set of genes from three gene-prediction programmes that exhibited weaker, but still significant, evidence that they may be expressed. Conservative criteria, requiring at least two lines of evidence, were used to define a set of 26,383 genes with good confidence.

Gene Annotation

Our cells are able to discern within the billions of base pairs of the genomic DNA the signals for initiating transcription and for splicing together exons separated by a few or hundreds of thousands of base pairs. The first step in characterising the genome is to define the structure of each gene and each transcription unit. The number of protein-coding genes in mammals has been controversial from the outset. Initial estimates based on reassociation data placed it between 30,000 to 40, 000, whereas later estimates from the brain were >100, 000.

More recent data from both the corporate and public sectors, based on extrapolations from EST, CpG island, and transcript density-based extrapolations, have not reduced this variance. The highest recent number of 142,634 genes emanates from a report from incyte Pharmaceuticals, and is based on a combination of EST data and the association of ESTs with CpG islands.

In stark contrast are three quite different, and much lower estimates: one of 35,000 genes derived with genome-wide EST data and sampling procedures in conjunction with chromosome 22 data; another of 28,000 to 34,000 genes derived with a comparative methodology involving sequence conservation between humans and the puffer fish *Tetraodon nigroviridis*; and a figure of 35,000 genes, which was derived simply by extrapolating from the density of 770 known and predicted genes in the 67 Mbp of chromosomes 21 and 22, to the approximately 3-Gbp euchromatic genome.

Computational identification of transcriptional units in genomic DNA sequence can be divided into two phases. The first is to partition the sequence into segments that are likely to correspond to individual genes. This is not trivial and is a weakness of most *de novo* gene-finding algorithms. It is also critical to determining the number of genes in the human gene inventory.

The second challenge is to construct a gene model that reflects the probable structure of the transcripts encoded in the region. This can be done with reasonable accuracy when a full-length cDNA has been sequenced or a highly homologous protein sequence is known. *De novo* gene prediction, although less accurate, is the only way to find genes that are not represented by homologous proteins or ESTs.

The rule-based expert system, Otto is used identify and characterise genes in the human genome. Otto attempts to simulate in software the process that a human annotator uses to identify a gene and refine its structure. In the process of annotating a region of the genome, a human curator examines the evidence provided by the computational pipeline and examines how various types of evidence relate to one another.

A curator puts different levels of confidence in different types of evidence and looks for certain patterns of evidence to support gene annotation. For example, a curator may examine homology to a number of ESTs and evaluate whether or not they can be connected into a longer, virtual m-RNA. The curator would also evaluate the strength of the similarity and the contiguity of the match, in essence asking whether any ESTs cross splice-junctions and whether the edges of putative exons have consensus splice sites. The *Drosophila* genome get annotated by this kind of manual annotation process.

The observed evidence to a gene annotation get promoted by the Otto system in one of two ways. First, if the evidence includes a high-quality match to the sequence of a known gene, then Otto can promote this to a gene annotation. In the second method, Otto evaluates a broad spectrum of evidence and determines if this evidence is adequate to support promotion to a gene annotation.

Gene boundaries are predicted on the basis of examination of sets of overlapping protein and EST matches generated by a computational pipeline. This pipeline searches the scaffold sequences against protein, EST, and genome-sequence databases to define regions of sequence similarity and runs three *de novo* gene-prediction programmes. To identify likely gene boundaries, regions of the genome were partitioned by Otto on the basis of sequence matches identified by BLAST.

Each of the database sequences matched in the region under analysis was compared by an algorithm that takes into account both coordinates of the matching sequence, as well as the sequence type. The results were used to group the matches into bins of related sequences that may define a gene and identify gene boundaries. During this process, multiple hits to the same region were collapsed to a coherent set of data by tracking the coverage of a region.

For example, if a group of bases was represented by multiple overlapping ESTs, the union of these regions matched by the set of ESTs on the scaffold was marked as being supported by EST evidence. This resulted in a series of 'gene bins', each of which was believed to contain a single gene. One weakness of this initial implementation of the algorithm was in predicting gene boundaries in regions of tandemly duplicated genes. Gene clusters frequently resulted in homologous neighbouring genes being joined together, resulting in an annotation that artificially concatenated these gene models.

Known genes, those with exact matches of a full-length cDNA sequence to the genome were identified, and the region corresponding to the cDNA was annotated as a predicted transcript. A subset of the curated human gene set RefSeq from the National Centre for Biotechnology information (NCBI) was included as a data set searched in the computational pipeline. If a RetSeq transcript matched the genome assembly for at least 50 per cent of its length at >92 per cent identity, then the SIM4 alignment of the RefSeq transcript to the region of the genome under analysis was promoted to the status of an Otto annotation.

Since the genome sequence has gaps and sequence errors such as frameshifts, it was not always possible to predict a transcript that agrees precisely with the experimentally determined cDNA sequence. A total of 6, 538 genes in HGP's inventory were identified and transcripts predicted in this way.

Part of the Otto system that uses the sequence similarity information to predict a transcript analysing regions that have a substantial amount of sequence similarity, but do not match known genes. Otto evaluates evidence generated by the computational pipeline, corresponding to conservation between mouse and human genomic DNA, similarity to human transcript (ESTs and cDNAs), similarity to rodent transcripts (ESTs and cDNAs) and similarity of the translation of human genomic DNA to known proteins to predict potential genes in the human genome.

The sequence from the region of genomic DNA contained in a gene bin was extracted, and the subsequences supported by any homology evidence were marked. The other bases in the region, those not covered by any homology evidence, were replaced by N's. This sequence segment, with high confidence regions represented by the consensus genomic sequence and the remainder represented by N's, was then evaluated by Genscan to see if a consistent gene model could be generated.

This procedure simplified the gene-prediction task by first establishing the boundary for the gene, and by eliminating regions with no supporting evidence. If Genscan returned a plausible gene model, it was further evaluated before being promoted to an Otto annotation. The final Genscan predictions were often quite different from the prediction that Genscan returned on the same region of native genomic sequence.

The next step was to compare each predicted transcript with the homology-based evidence that was used in previous steps to evaluate the depth of evidence for each exon in the prediction. Internal exons were considered to be supported if they were covered by homology evidence to within ± bases of their edges. For first and last exons, the internal edge was required to be within 10 bases, but the external edge was allowed greater latitude to allow for 5' and 3' untranslated regions (UTRs).

To be retained a prediction for a multi-exon gene must have evidence such that the total number of 'hits' divided by the number of exons in the prediction must be >0.66 or must correspond to a RefSeq sequence. A single-exon gene must be covered by at least three supporting hits and these must cover the complete predicted open reading frame. For a single-exon gene, HGP also required that the Genscan prediction include both a start and a stop codon.

Gene models that did not meet these criteria were disregarded, and those that passed were promoted to Otto predictions. Homology-based Otto predictions do not contain 3' and 5' untranslated sequence. Although three *de novo* gene-finding programmes were run as part of the computational analysis, the results of these programmes were not directly used in making the Otto predictions.

Otto Validation

The Otto homology-based process and the method that Otto uses to define the structures of known genes, validated by comparing transcripts predicted by Otto with their corresponding transcript from a set of 4512 RefSeq transcripts for which there was a unique SIM4 alignment. In order to evaluate the relative performance of Otto and Genscan, HGP made three comparisons.

The first involved a determination of the accuracy of gene models predicted by Otto with only homology data other than the corresponding RefSeq sequence. HGP measured the sensitivity and specificity. Second, HGP examined the sensitivity and specificity of the Otto predictions that were made solely with the RefSeq sequence, which is the process that Otto uses to annotate known genes. And third, HGP determined the accuracy of the Genscan predictions corresponding to these RefSeq sequences.

The alignment method was the most accurate, and Otto-homology performed better than Genscan by both criteria. Thus. 6.1 per cent of true RefSeq nucleotides were not represented in the Ottorefseq annotations and 2.7 per cent of the nucleotides in the Otto-RefSeq transcripts were not contained in the original RefSeq transcripts. The discrepancies could come from legitimate differences between the Celera assembly and the RefSeq transcript due to polymorphisms, incomplete or incorrect data in the Celera assembly, errors introduced by SIM4 during the alignment process, or the presence of alternatively spliced forms in the data set used for the comparisons. The absence of experimental evidence for intervening exons may inadvertently result in a set of exons that cannot be spliced together to give rise to a transcript. In such cases, Otto may 'split genes' when in fact all the evidence should be combined into a single transcript.

Gene Number

The Otto system is quite conservative so that HGP used a different gene-prediction strategy in regions where the homology evidence was less strong. Here the results of *de novo* gene predictions were used. For these genes, HGP insisted that a predicted transcript have at least two of the following types of evidence to be included in the gene set for further analysis: protein human EST, rodent EST, or mouse genome fragment matches. This final class of predicted genes is a subset of the predictions made by the three gene-finding programmes that were used in the computational pipeline. For these, there was not sufficient sequence similarity information for Otto to attempt to predict a gene structure.

The three *de novo* gene-finding programmes resulted in about 155,695 predictions, of which ~76,410 were non-redundant. Of these, 57,935 did not overlap known genes or predictions made by Otto. Only 21,350 of the gene predictions that did not overlap Otto predictions were partially supported by at least one type of sequence similarity evidence, and 8, 619 were partially supported by two types of evidence. The sum of this number (21,350) and the number of Otto annotations (17,764), 39,114, is near the upper limit for the human gene complement. If the requirement for other supporting evidence is made more stringent, this number drops rapidly so that demanding two types of evidence reduces the total gene number to 26,383 and demanding three types reduces it to ~23,000. Requiring that a prediction be supported by all four categories of evidence is too stringent because it would remove protein encoding genes.

In an attempt to identify genes that were not found by the auto-annotation process or any of the *de novo* gene finders, HGP examined regions outside of gene predictions that were similar to the EST sequence, and where the EST matched the genomic sequence across a splice junction. After correcting for potential 3' UTRs of predicted genes, about 2, 500 such regions remained. Addition of a requirement for at least one of the following evidence types: homology to mouse genomic sequence fragments, rodent ESTs, or cDNAs—or similarity to a known protein reduced this number to 1, 010. Adding this to the numbers previously calculated gives estimates of about 40, 000, 27, 000, and 24, 000 potential genes in the human genome, depending on the stringency of evidence considered.

Transcripts encoded by a set of 26, 383 genes includes the 6, 538 genes predicted by Otto on the basis of matches to known genes, 11, 226 transcripts predicted by Otto based on homology evidence, and 8, 619 from the subset of transcripts from de novo gene-prediction programmes that have two types of supporting evidence. These are a very preliminary set of annotations and are subject to all the limitations of an automated process. Considerable refinement is still necessary to improve the accuracy of these transcript predictions.

All the predictions and descriptions of genes and the associated evidence that HGP present are the product of completely computational processes, not expert curation. HGP has attempted to enumerate the genes in the human genome in such a way that HGP has different levels of confidence based on the amount of supporting evidence: known genes, genes with good protein or EST homology evidence and *de novo* gene predictions confirmed by modest homology evidence.

Human Gene Transcripts

HGP estimated the average span for a typical gene in the human DNA sequence as about 27,894 bases. This is based on the average span covered by RefSeq transcripts, used because it represents the highest confidence set. The set of transcripts promoted to gene annotations varies in a number of ways. Transcripts predicted by Otto tend to be longer, having on average about 7.8 exons, whereas those promoted from gene-prediction programmes average about 3.7 exons. Largest number of exons identified in a transcript is 234 in titin m-RNA.

GENOME STRUCTURE

Cytogenetic Maps

The most obvious element of the structure of the genome is the banding pattern produced by Giemsa stain. About 17 per cent to

20 per cent of the human chromosome complement consists of C-bands, or constitutive heterochromatin. Much of this heterochromatin is highly polymorphic and consists of different families of alpha satellite DNAs with various higher order repeat structures.

Many chromosomes have complex inter- and intra-chromosomal duplications present in pericentromeric regions. About 5 per cent of the sequence reads were identified as alpha satellite sequences; these were not included in the assembly. The remaining ~80 per cent of the genome, the euchromatic component, is divisible into G-, R-, and T-bands. These cytogenetic bands have been presumed to differ in their nucleotide composition and gene density, although HGP has been unable to determine precise band boundaries at the molecular level. T-bands are the most G + C- and gene-rich, and G-bands are G + C-poor.

Bernardi has offered a description of the euchromatin at the molecular level as long stretches of DNA of differing base composition, termed isochores (denoted L, H1, H2, and H3), which are >300 kbp in length. Bernardi defined the L (light) isochores as G + C-poor (< 43 per cent), whereas the H (heavy) isochores fall into three G + C-rich classes representing 24, 8, and 5 per cent of the genome. Gene concentration has been claimed to be very low in the L isochores and 20-fold more enriched in the H2 and H3 isochores.

By examining contiguous 50-kbp windows of G + C content across the assembly, HGP found that regions of G + C content > 48 per cent (H3 isochores) averaged 273.9 kbp in length, those with G + C content between 43 and 48 per cent (H1 + H2 isochores) averaged 202.8 kbp in length, and the average span of regions with <43 per cent (L isochores) was 1, 078.6 kbp. The correlation between G + C content and gene density was also examined in 50-kbp windows along the assembled sequence.

The density of genes was greater in regions of high G + C than in regions of low G + C content. However, the correlation between G + C content and gene density was not as skewed as previously predicted. A higher proportion of genes were located in the G+C-poor regions than had been expected. Chromosomes 17, 19, and 22, which have a disproportionate number of H3-containing bands, had the highest gene density. Conversely, of the chromosomes that found to have the lowest gene density, X, 4, 18, 13, and Y, also have the fewest H3 bands. Chromosome 15, which also has few H3 bands, did not have a particularly low gene density in HGP analysis. In addition, chromosome 8, which found to have a low gene density, does not appear to be unusual in its H3 banding.

The human genome does indeed contain deserts, or large, gene-poor regions. If we define a desert as a region >500 kbp without a gene, then we see that 605 Mbp, or about 20 per cent of the genome, is in deserts. These are not uniformly distributed over the various chromosomes. Gene-rich chromosomes 17, 19, and 22 have only about 12 per cent of their collective 171 Mbp in deserts, whereas gene-poor chromosomes 4, 13, 18, and X have 27.5 per cent of their 492 Mbp in deserts.

Linkage Map

Linkage maps are widely used in the study of the inheritance of traits and in the positional cloning of genes. The distance metric, centimorgans (CM), is based on the recombination rate between homologous chromosomes during meiosis. In general, the rate of recombination in females is greater than that in males, and this degree of map expansion is not uniform across the genome.

One of the opportunities enabled by a nearly complete genome sequence is to produce the ultimate physical map, and to fully analyse its correspondence with two other maps that have been widely used in genome and genetic analysis: the linkage map and the cytogenetic map. This would close the loop between the mapping and sequencing phases of the genome project. HGP mapped the location of the markers that constitute the Genethon linkage map to the genome. Higher rates of recombination in the telomeric region of the chromosomes have been previously documented. From this mapping result, there is a difference of 4.99 between lowest rates and highest rates and the largest difference of 4.4 between males and females. This indicates that the variability in recombination rates among regions of the genome exceeds the differences in recombination rates between males and females.

The human genome has recombination hotspots, where recombination rates vary fivefold or more over a space of 1 kbp, so the picture one gets of the magnitude of variability in recombination rate will depend on the size of the window examined. Unfortunately, too few meiotic crossovers have occurred in centre *d'etude du* Polymorphism Humain (CEPH) and other reference families to provide a resolution any finer than about three Mbp.

The next challenge will be to determine a sequence basis of recombination at the chromosomal level. An accurate predictor for the rate for variation in recombination rates between any pair of markers would be extremely useful in designing markers to narrow a region of linkage.

CpG Islands and Genes

CpG islands are believed to preferentially occur at the transcriptional start of genes. Actually, they are stretches of unmethylated DNA with a higher frequency of CpG dinucleotides when compared with the entire genome. Most housekeeping genes have CpG islands at the 5' end of the transcript.

Experimental evidence indicates that CpG island methylation is correlated with gene inactivation and has been shown to be important during gene imprinting and tissue-specific gene expression. Experimental methods have been used that resulted in an estimate of 30,000 to 45,000 CpG islands in the human genome and an estimate of 499 CpG islands on human chromosome 22.

Larsen *et al.*, and Gardiner-Garden and Frommer used a computational method to identify CpG islands and defined them as regions of DNA of >200 bp that have a G + C content of >50 per cent and a ratio of observed versus expected frequency of CG dinucleotide ≥ 0.6. It is difficult to make a direct comparison of experimental definitions of CpG islands with computational definitions because computational methods do not consider the methylation state of cytosine and experimental methods do not directly select regions of high G + C content. However, we can determine the correlation of CpG island with gene starts, given a set of annotated genomic transcripts and the whole genome sequence.

HGP has analysed the publicly available annotation of chromosome 22, as well as using the entire human genome in the assembly and the computationally annotated genes. A variation of the CpG island computation was compared with Larsen *et al*. The main differences are that HGP used a sliding window of 200 bp, consecutive windows are merged only if they overlap, and recomputed the CpG value upon merging, thus rejecting any potential island if it scores less than the threshold.

HGP used two different thresholds of CG dinucleotide likelihood ratio to compute various CpG statistics. Besides using the original threshold of 0.6 (method 1), HGP used a higher threshold of CG dinucleotide likelihood ratio of 0.8 (method 2), which results in the number of CpG islands on chromosome 22 close to the number of annotated genes on this chromosome.

CpG islands computed with method 1 predicted only 2.6 per cent of the CSA sequence as CpG, but 40 per cent of the gene starts (start codons) are contained inside a CpG island. The observed average closest CpG islands are smaller than the corresponding expected distances, confirming an association between CpG island and the first exon. HGP also looked at the distribution of CpG island nucleotides among various sequence classes such as intergenic regions, introns, exons, and first exons. HGP computed the likelihood score for each sequence class as the ratio of the observed fraction of CpG island nucleotides in that sequence class and the expected fraction of CpG island nucleotides in that sequence class. The result of applying method 1 on CSA were scores of 0.89 for intergenic region, 1.2 for intron, 5.86 for exon, and 13.2 for first exon. The same trend was also found for chromosome 22 and after the application of a higher threshold on both data sets.

Genome-wide Repetitive Elements

HGP observed about 35 per cent of the genome in repeatitive DNA classes. Repetitive sequence may be underrepresented in the Celera assembly as a result of incomplete repeat resolution. About eight per cent of the scaffold length is in gaps, and HGP expect that much of this is repetitive sequence. Chromosome 19 has the highest repeat density, as well as the highest gene density. Of interest, among the different classes of repeat elements, HGP observe a clear association of alu elements and gene density, which was not observed between LINEs and gene density.

GENOME EVOLUTION

The dynamic nature of genome evolution can be captured at several levels. These include gene duplications mediated by RNA intermediates (retrotransposition) and segmental genomic duplications. This section documents the genome-wide occurrence of retrotransposition events generating functional (intronless paralogs) or inactive genes (pseudogenes).

Retrotransposition

Processed m-RNA transcripts' retrotransposition into the genome results in functional genes, called intronless paralogs, or inactivated genes (pseudogenes). A paralog refers to a gene that appears in more than one copy in a given organism as a result of a duplication event. Cataloguing these evolutionary events on the genomic landscape is of value in understanding the functional consequences of such gene-duplication events in cellular biology.

Identification of conserved intronless paralogs in the mouse or other mammalian genomes should provide the basis for capturing the evolutionary chronology of these transposition events and provide insights into gene loss. A set of proteins corresponding to all 901 Otto-predicted, single-exon genes were subjected to BLAST analysis against the proteins encoded by the remaining multiexon predicted transcripts. Using homology criteria of 70 per cent sequence identity over 90 per cent of the length, HGP identified 298 instances of single-to multi-exon correspondence.

Of these 298 sequences, 97 were represented in the GenBank data set of experimentally validated full-length genes at the stringency specified and were verified by manual inspection. These 97 cases may represent intronless paralogs of known genes. Most of these are flanked by direct repeat sequences, although the precise nature of these repeats remains to be determined.

Retrotransposition may serve as a mechanism used to escape X-chromosomal inactivation. Several cases of retrotransposition from a single source chromosome to multiple target chromosomes. Interesting examples include the retrotransposition of a five exon-containing ribosomal protein L21 gene on chromosome 13 onto chromosomes 1, 3, 4, 7, 10, and 14, respectively. The size of the source genes can also show variability. The largest example is the 31-exon diacylglycerol kinase zeta gene on chromosome 11 that has an intronless paralog on chromosome 13. Regardless of route, retrotransposition with subsequent gene changes in coding or non-coding regions that lead to different functions or expression patterns, represents a key route to providing an enhanced functional repertoire in mammals.

The preliminary set of retrotransposed intronless paralogs contains a clear over-representation of genes involved in translational processes and nuclear regulation, as well as metabolic and regulatory enzymes. EST matches specific to a subset of intronless paralogs suggest expression of these intronless paralogs. Differences in the upstream regulatory sequences between the source genes and their intronless paralogs could account for differences in tissue-specific gene expression.

Pseudogenes

Pseudogenes are nonfunctional copies that are very similar to normal genes but that has been altered slightly so that it is not expressed. HGP developed a method for the preliminary analysis of processed pseudogenes in the human genome as a starting point in elucidating the ongoing evolutionary forces that account for gene inactivation. The general structural characteristics of these processed pseudogenes include the complete lack of intervening sequences found in the functional counterparts, a poly(a) tract at the 3′ end, and direct repeats flanking the pseudogene sequence.

Processed pseudogenes occur as a result of retrotransposition, whereas unprocessed pseudogenes arise from segmental genome, duplication. HGP searched the complete set of Otto-predicted transcripts against the genomic sequence by means of BLAST. Genomic regions corresponding to all Otto-predicted transcripts were excluded from this analysis. HGP identified 2, 909 regions matching with greater than 70 per cent identity over at least 70 per cent of the length of the transcripts that likely represent processed pseudogenes. HGP looked for correlations between structural elements and the propensity for retrotransposition in the human genome. QC content and transcript length were compared between the genes with processed pseudogenes versus the remainder of the predicted gene set.

Transcripts that give rise to processed pseudogenes have shorter average transcript length as compared with genes for which no pseudogene was detected. The overall GC content did not show any significant difference, contrary to a recent report. There is a clear trend in gene families that are present as processed pseudogenes. These include ribosomal proteins, lamin receptors, translation elongation factor alpha, and HMG-non-histone proteins. The increased occurrence of retrotransposition among genes involved in translation and nuclear regulation may reflect an increased transcriptional activity of these genes.

Gene Duplication

HGP developed a graph-theoretic algorithm, called Lek, for grouping the predicted human protein set into protein families. The complete clusters that result from the Lek clustering provide one basis for comparing the role of whole-genome or chromosomal duplication in protein family expansion as opposed to other means, such as tandem duplication.

Since each complete cluster represents a closed and certain island of homology, and because Lek is capable of simultaneously clustering protein complements of several organisms, the number of proteins contributed by each organism to a complete cluster can be predicted with confidence depending on the quality of the annotation of each genome.

The variance of each organism's contribution to each cluster can then be calculated, allowing an assessment of the relative importance of large-scale duplication versus smaller-scale, organism-specific expansion and contraction of protein families, presumably as a result of natural selection operating on individual protein families within an organism.

The large variance in the relative numbers of human as compared with *D. melanogaster* and *Caenorhabditis elegans* proteins in complete clusters may be explained by multiple events of relative expansions in gene families in each of the three animal genomes. Such expansions would give rise to the distribution that shows a peak at 1: 1 in the ratio for human-worm or human-fly clusters with the slope spread covering both human and fly/ worm predominance.

Furthermore, there are nearly as many clusters where worm and fly proteins predominate despite the larger numbers of proteins in the human. At face value, this analysis suggests that natural selection acting on individual protein families has been a major force driving the expansion of at least some elements of the human protein set. However, the difference between an ancient whole-genome duplication followed by loss, versus piecemeal duplication, cannot be easily distinguished.

To search large-scale duplications in the human genome, two independent methods were used. One is a protein family-based method that identified highly conserved blocks of duplication and the other is a comprehensive method for identifying all interchromosomal block duplications.

The protein family-based method is based on the idea of searching for blocks of highly conserved homologous proteins that occur in more than one location on the genome. For this comparison, two genes were considered equivalent if their protein products were determined to be in the same family and the same complete Lek cluster.

Initially, each chromosome was represented as a string of genes ordered by the start codons for predicted genes along the chromosome. HGP considered the two strands as a single string, because local inversions are relatively common events relative to large-scale duplications. Each gene was indexed according to the protein family and Lek complete cluster.

All pairs of indexed gene strings were then aligned in both the forward and reverse directions with the Smith-Waterman algorithm. A match between two proteins of the same Lek complete cluster was given a score of 10 and a mismatch –10, with gap open and extend penalties of –4 and –1. With these parameters, 19 conserved interchromosomal blocks of duplication were observed. The detection of only a relatively small number of block duplications was a consequence of using an intrinsically conservative method grounded in the conservative constraints of the complete Lek clusters.

In a more comprehensive approach, HGP aligned all chromosomes directly with one another using an algorithm based on the MuMmer system. This alignment method uses a suffix tree data structure and a linear-time algorithm to align long sequences very rapidly; for example, two chromosomes of 100 Mbp can be aligned in less than 20 min with four gigabytes of memory. This procedure was used recently to identify numerous large-scale segmental duplications among the five chromosomes of *A. thaliana*; in that organism, the method revealed that 60 per cent of the genome (66 Mbp) is covered by 24 very large duplicated segments.

For *arabidopsis,* a DNA-based alignment was·sufficient to reveal the segmental duplications between chromosomes; in the human genome. DNA alignments at the whole-chromosome level are insufficiently sensitive. Therefore, a modified procedure was developed.

First, all 26,588 proteins were concatenated end-to-end in order as they occur along each of the 24 chromosomes, irrespective of strand location. The concatenated protein set was then aligned against each chromosome, by the MuMmer algorithm. The resulting matches were clustered to extract all sets of three or more protein matches that occur in close proximity on two different chromosomes; these represent the candidate segmental duplications.

A series of filters were developed and applied to remove likely false-positives from this set; for example, small blocks that were spread across m-proteins were removed. To refine the filtering methods, a shuffled protein set was first created by taking the 26,588 proteins randomising their order, and then partitioning them into 24 shuffled chromosomes, each containing the same number of proteins as the true genome. This shuffled protein set has the identical composition to the real genome; in particular, every protein and every domain appears the same numbe. of times.

The complete algorithm was then applied to both the real and the shuffled data, with the results on the shuffled data being used to estimate the false-positive rate. The algorithm after filtering yielded 10,3 10 gene pairs in 1, 077 duplicated blocks containing 3, 522 distinct genes tandemly duplicated expansions in many of the blocks explain the excess of gene pairs to distinct genes. In the shuffled data, by contrast, only 370 gene pairs were found, giving a false-positive estimate of 3.6 per cent.

The most likely explanation for the 1, 077 block duplications is ancient segmental duplications. In many cases, the order of the proteins has been shuffled, although proximity is preserved out of the 1, 077 blocks, 159 contain only three genes, 137 contain four genes, and 781 contain five or more genes.

The duplications are ubiquitous in the genome. It displays many relatively small chromosomal stretches, with one-to-many duplication relationships that are graphically striking. One such example captured by the analysis is the well-documented olfactory receptor (OR) family, which is scattered in blocks throughout the genome and which has been analysed for genome development reconstructions at several evolutionary stages.

Some chromosomes, such as chromosome 2, contain many more detected large-scale duplications than others. Indeed, one of the largest duplicated segments is a large block of 33 proteins on chromosome 2, spread among eight smaller blocks in 2p, that aligns to a paralogous set on chromosome 14, with one rearrangement. The proteins are not contiguous but span a region containing 97 proteins on chromosome 2 and 332 proteins on chromosome 14. The likelihood of observing this many duplicated proteins by chance, even over a span of this length, is 2.3× 10.

This duplicated set spans 20 Mbp on chromosome 2 and 63 Mbp on chromosome 14, over 70 per cent of the latter chromosome. Chromosome 2 also contains a block duplication that is nearly as large, which is shared by chromosome arm 2q and chromosome 12. This duplication incorporates two of the four known Hox gene clusters, but considerably expands the extent of the duplications proximally and distally on the pair of chromosome arms. This breadth of duplication is also seen on the two chromosomes carrying the other two Hox clusters. An additional large duplication, between chromosomes 18 and 20, serves as a good example to illustrate some of the features common to many of the other observed large duplications contains 64 detected ordered intrachromosomal pairs of homologous genes. After discounting a 40-Mb stretch of chromosome 18 free of matches to chromosome 20, which is likely to represent a large insert, the full duplication segment covers 36 Mb on chromosome 18 and 28 Mb on chromosome 20.

The duplication segment spans nearly half of each chromosome's net length. The most likely scenario is that the whole span of this region was duplicated as a single very large block, followed by shuffling owing to smaller scale rearrangements. As such, at least four subsequent rearrangements would need to be invoked to explain the relative insertions and inversions seen in the duplicated segment interval. The 64 protein pairs in this alignment occur among 217 protein assignments on chromosome 18, and among 322 protein assignments on chromosome 20, for a density of involved proteins of 20 to 30 per cent. This is consistent with an ancient large-scale duplication followed by subsequent gene loss on one or both chromosomes.

Loss of just one member of a gene pair subsequent to the duplication would result in a failure to score a gene pair in the block, less than 50 per cent gene loss on the chromosomes would lead to the duplication density observed here. As an independent verification of the significance of the alignments detected, a substantial number of the pairs of aligning proteins in this duplication, including some of those annotated, are those populating small Lek complete clusters. This indicates that they are members of very small families of paralogs; their relative scarcity within the genome validates the uniqueness and robust nature of their alignments.

Among many of the large-scale duplications, two additional qualitative features were observed.

First, several proteins with disease associations, with Online Mendelian Inheritance in Man (OMIM) assignments, are members of duplicated segments. HGP has also observed a few instances where paralogs on both duplicated segments are associated with similar disease conditions. Notable among these genes are proteins involved in hemostasis that are associated with bleeding disorders, transcriptional regulators like the homeobox proteins associated with developmental disorders and potassium channels associated with cardiovascular conduction abnormalities. For each of these disease genes, closer study of the paralogous genes in the duplicated segment may reveal new insights into disease causation, with further investigation needed to determine whether they might be involved in the same or similar genetic diseases.

Second, although there is a conserved number of proteins and coding exons predicted for specific large duplicated spans within the chromosome 18 to 20 alignment, the genomic DNA of chromosome 18 in these specific spans is in some cases more than 10-fold longer than the corresponding chromosome 20 DNA. This selective accretion of non-coding DNA on one of a pair of duplicated chromosome regions was observed in many compared regions.

Large-scale ancient segmental duplication in fact best explains many of the blocks detected by this genome-wide analysis. The regions of human chromosomes involved in the large-scale duplications expanded upon above are each syntenic to a distinct mouse chromosomal region. The corresponding mouse chromosomal regions are much more similar in sequence conservation, and even in order, to their human synteny partners than the human duplication regions are to each other. Further, the corresponding mouse chromosomal regions each bear a significant proportion of genes orthologous to the human genes on which the human duplication assignments were made. On the basis of these factors, the corresponding mouse chromosomal spans, at coarse resolution, appear to be products of the same large-scale duplications observed in humans.

Although further detailed analysis must be carried out once a more complete genome is assembled for mouse, the underlying large duplications appear to predate the two species divergence. This dates the duplications at the latest, before divergence of the primate and rodent lineages. This date can be further refined upon examination of the synteny between human chromosomes and those of chicken, pufferfish or zebrafish. The only substantial syntenic stretches mapped in these species corresponding to both pairs of human duplications are restricted to the Hox cluster regions. When the synteny of these regions to human chromosomes is extended with further mapping, the ages of the nearly chromosome-length duplications seen in humans are likely to be dated to the root of vertebrate divergence.

The extent of segmental duplications raises the question of whether an ancient whole-genome duplication event is the underlying explanation for the numerous duplicated regions. The duplications have undergone many deletions and subsequent rearrangements; these events make it difficult to distinguish between a whole-genome duplication and multiple smaller events. Further analysis, focused especially on comparing the estimated ages of all the block duplications, derived partially from interspecies genome comparisons, will be necessary to determine which of these two hypotheses is more likely. Comparisons of genomes of different vertebrates, and even cross-phyla genome comparisons, will allow for the deconvolution of duplications to eventually reveal the stagewise history of our genome.

GENOME SEQUENCE VARIATIONS

To identify single-nucleotide polymorphisms (SNPs) computational methods were used. The SNP rate between two chromosomes was 1 per 1, 200 to 1, 500 bp. SNPs are distributed nonrandomly throughout the genome. Only a very small proportion of all SNPs (<1 per cent) potentially impact protein function based on the functional analysis of SNPs that affect the predicted coding regions. This results in an estimate that only thousands, not millions, of genetic variations may contribute to the structural diversity of human proteins.

Analysis of sequence variation in DNA can enable us to discover the genetic basis for variation in health among human beings. Whole-genome shotgun sequencing is a particularly effective method for detecting sequence variation in tandem with whole-genome assembly. In addition, HGP compared the distribution and attributes of SNPs ascertained by three other methods:

i) alignment of the Celera consensus sequence to the PFP assembly,
ii) overlap of high-quality reads of genomic sequence and
iii) reduced representation shotgun sequencing.

These above data were consistent in showing an overall nucleotide diversity of $\sim 8 \times 10^{-4}$, marked heterogeneity across the genome in SNP density, and an overwhelming preponderance of noncoding variation that produces no change in expressed proteins.

Methods of SNP discovery make full use of sequence depth and quality at every site and quantitatively control the rate of false-positive and false-negative calls with an explicit sampling model. Comparison of consensus sequences in the absence of these details necessitated a more ad hoc approach. All sequence differences between the two consensus sequences were identified; these were then filtered to reduce the contribution of sequencing errors and misassembly.

As a measure of the effectiveness of the filtering step, HGP monitored the ratio of transition and transversion substitutions, because a 2: 1 ratio has been well documented as typical in mammalian evolution and in human SNPs. The filtering steps consisted of removing variants where the quality score in the Celera consensus was less than 30 and where the density of variants was greater than five in 400 bp. These filters resulted in shifting the transition-to-transversion ratio from 1.57: 1 to 1.89: 1. When applied to 2.3 Gbp of alignments between the Celera and PFP consensus sequences, these filters resulted in identification of 2,104,820 putative SNPs from a total of 2,778,474 substitution differences.

Additional SNPs, including 2,536,021 from dbSNP and 13,150 from Human Gene Mutation Database (HGMD), were mapped on the Celera consensus sequence by a sequence similarity search with the programme PowerBlast. The two largest data sets in dbSNP are the Kwok and TSC sets, with 47 per cent and 25 per cent of the dbSNP records. Low-quality alignments with partial coverage of the dbSNP sequence and alignments that had less than 98 per cent sequence identity between the Celera sequence and the dbSNP flanking sequence were eliminated dbSNP sequences mapping to multiple locations on the Celera genome were discarded.

A total of 2,336,935 dbSNP variants were mapped to 1,223,038 unique locations on the Celera sequence, implying considerable redundancy in dbSNP. SNPs in the TSC set mapped to 585,811 unique genomic locations, and SNPs in the Kwok set mapped to 438,032 unique locations. The combined unique SNPs counts used in this analysis, including Celera-PFP, TSC, and Kwok, is 2,737,668.

A substantial fraction of SNPs identified by one of these methods was also found by another method. The very high overlap between the Kwok and Celera-PFP SNPs may be due in part to the use by Kwok of sequences that went into the PFP assembly. The unusually low overlap between the Kwok and TSC sets is due to their being the smallest two sets. In addition, 24.5 per cent of the Celera-PFP SNPs overlap with SNPs derived from the Celera genome sequences. SNP validation in population samples is an expensive and laborious process, so confirmation on multiple data sets may provide an efficient initial validation *'in silico'*. One means of assessing whether the three sets of SNPs provide the same picture of human variation is to tally the frequencies of the six possible base changes in each set of SNPs.

Previous measures of nucleotide diversity were mostly derived from small-scale analysis on candidate genes, and HGP's analysis with all three data sets validates the previous observations at the whole-genome scale. There is remarkable homogeneity between the SNPs found in the Kwok set, the TSC set, and in whole-genome shotgun in this substitution pattern.

Compared with the rest of the data sets, Celera-PFP deviates slightly from the 2: 1 transition-to-transversion ratio observed in the other SNP sets. This result is not unexpected, because some fraction of the computationally identified SNPs in the Celera-PFP comparison may in fact be sequence errors. A 2: 1 transition: transversion ratio for the bona fide SNPs would be obtained if one assumed that 15 per cent of the sequence differences in the Celera-PFP set were a result of sequence errors.

Nucleotide Diversity

Nucleotide diversity is a measure of per-site heterozygosity, quantifying the probability that a pair of chromosomes drawn from the population will differ at a nucleotide site. In order to calculate nucleotide diversity for each chromosome, we need to know the number of nucleotide sites that were surveyed for variation, and in methods like reduced representation sequencing, we need to know the sequence quality and the depth of coverage at each site. These data are not readily available, so we could not estimate nucleotide diversity from the TSC effort.

Estimation of nucleotide diversity from high-quality sequence overlaps should be possible, but again, more information is needed on the details of all the alignments. Estimation of nucleotide diversity from a shotgun assembly entails calculating for each column of the multialigmnent, the probability that two or more distinct alleles are present, and the probability of detecting a SNP if in fact the alleles have different sequence.

The greater the depth of coverage and the higher the sequence quality, the higher is the chance of successfully detecting a SNP. Even after correcting for variation in coverage, the nucleotide diversity appeared to vary across autosomes. The significance of this heterogeneity was tested by analysis of variance, with estimates of π for 100-kbp windows to estimate variability within chromosomes. Average diversity for the autosomes estimated from the Celera-PFP comparison was 8.94 $\times 10^{-4}$.

Nucleotide diversity on the X chromosome was 6.54×10^{-4}. The X is expected to be less variable than autosomes, because for every four copies of autosomes in the population, there are only three X chromosomes, and this smaller effective population size means that random drift will more rapidly remove variation from the X. Having ascertained nucleotide variation genome-wide, it appears that previous estimates of nucleotide diversity in humans based on samples of genes were reasonably accurate. Genome-wide, HGP's estimate of nucleotide diversity was 8.98×10^{-4} for the Celera-PFP alignment, and a published estimate averaged over 10 densely resequenced human genes was 8.00×10^{-4}.

If SNPs occur by random and independent mutations, then it would seem that there ought to be a Poisson distribution of numbers of SNPs in fragments of arbitrary constant size. The observed dispersion in the distribution of SNPs in 100-kbp fragments was far greater than predicted from a Poisson distribution. However, this simplistic model ignores the different recombination rates and population histories that exist in different regions of the genome.

Population genetics theory holds that we can account for this variation with a mathematical formulation called the neutral coalescent. Applying well-tested algorithms for simulating the neutral coalescent with recombination and using an effective population size of 10,000 and a per base recombination rate equal to the mutation rate, HGP generated a distribution of numbers of SNPs by this model as well. The observed distribution of SNPs has a much larger variance than either the Poisson model or the coalescent model and the difference is highly significant.

Attributes of the DNA sequence may affect the local density of SNPs, including the rate at which DNA polymerase makes errors and the efficacy of mismatch repair. One key factor that is likely to be associated with SNP density is the G + C content, in part because methylated cytosines in CpG dinucleotides tend to undergo deamination to form thymine, accounting for a nearly 10-fold increase in the mutation rate of CpGs over other dinucleotides. HGP tallied the GC content and nucleotide diversities in 100-kbp windows across the entire genome and found that the correlation between them was positive and highly significant.

To test homogeneity of SNP densities across functional classes, HGP partitioned sites into intergenic, 5′–UTR, exonic, intronic, and 3′–UTR for 10,239 known genes, derived from the NCBI RefSeq database and all human genes predicted from the Celera Otto annotation. In coding regions, SNPs were categorised as either silent, for those that do not change amino acid sequence, or missense, for those that change the protein product.

The ratio of missense to silent coding SNPs in Celera-PFP, TSC, and Kwok sets shows a markedly reduced frequency of missense variants compared with the neutral expectation, consistent with the elimination by natural selection of a fraction of the deleterious amino acid changes. These ratios are comparable to the missense-to-silent ratios of 0.88 and 1.17 found by Cargill *et al.,* and by Halushka *et al.* Similar results were observed in SNPs derived from Celera shotgun sequences.

It is striking how small is the fraction of SNPs that lead to potentially dysfunctional alterations in proteins. In the 10, 239 RefSeq genes, missense SNPs were only about 0. 12, 0.14, and 0. 17 per cent of the total SNP counts in Celera-PFP, TSC, and

Kwok SNPs, respectively. Nonconservative protein changes constitute an even smaller fraction of missense SNPs. Intergenic regions have been virtually unstudied.

The SNP rate was highest in introns and lowest in exons. The SNP rate was lower in intergenic regions than in introns, providing one of the first discriminators between these two classes of DNA. These SNP rates were confirmed in the Celera SNPs, which also exhibited a lower rate in exons than in introns, and in extragenic regions than in introns. Many of these intergenic SNPs will provide valuable information in the form of markers for linkage and association studies.

PROTEIN-CODING GENES IN THE HUMAN GENOME

Over 40 per cent of the predicted protein set in humans cannot be ascribed a molecular function by methods that assign proteins to known families. A protein domain-based analysis provides a detailed catalogue of the prominent differences in the human genome when compared with the fly and worm genomes. Prominent among these are domain expansions in proteins involved in developmental regulation and in cellular processes such as neuronal function, hemostasis, acquired immune response, and cytoskeletal complexity. The final enumeration of protein families and details of protein structure will rely on additional experimental work and comprehensive manual curation.

Two methods were used to analyse and classify the molecular functions of 26,588 predicted proteins that represent 26,383 gene predictions with at least two lines of evidence as described above. The first method was based on an analysis at the level of protein families, with both the publicly available Pfam database and Celera's Panther Classification (CPC). The second method was based on an analysis at the level of protein domains, with both the Pfam and SMART databases.

Both the gene predictions and functional assignments have been made by using computational tools, although the statistical models in Panther, Pfam, and SMART have been built, annotated, and reviewed by expert biologists. In the set of computationally predicted genes, HGP expect both false-positive predictions and false-negative predictions. HGP also expect errors in delimiting the boundaries of exons and genes. Similarly, in the automatic functional assignments, HGP also expect both false-positive and false-negative predictions.

The functional assignment protocol focuses on protein families that tend to be found across several organisms, or on families of known human genes. Therefore, HGP do not assign a function to many genes that are not in large families, even if the function is known. Unless otherwise specified, all enumeration of the genes in any given family or functional category was taken from the set of 26,588 predicted proteins, which were assigned functions by using statistical score cutoffs defined for models in Panther, Pfam, and SMART. For this initial examination of the predicted human protein set, three broad questions were asked:

i) What are the likely molecular functions of the predicted gene products, and how are these proteins categorised with current classification methods?

ii) What are the core functions that appear to be common across the animals?

iii) How does the human protein complement differ from that of other sequenced eukaryotes?

Molecular Functions

About 41 per cent of the gene products could not be classified from initial analysis and are termed proteins with unknown functions. Since the automatic classification methods treat only relatively large protein families, there are a number of unclassified sequences that do, in fact, have a known or predicted function. For the 60 per cent of the protein set that have automatic functional predictions, the specific protein functions have been placed into broad classes.

We focus here on molecular function in order to classify as many proteins as possible. In the analysis of the 12,731 additional low-confidence predicted genes, only 636 of these additional putative genes were assigned molecular functions by the automated methods. One-third of these 636 predicted genes represented endogenous retroviral proteins, further suggesting that the majority of these unknown-function genes are not real genes. Given that most of these additional 12,095 genes appear to be unique among the genomes sequenced to date, many may simply represent false-positive gene predictions.

The common molecular functions are the transcription factors and those involved in nucleic acid metabolism. Other functions that are highly represented in the human genome are the receptors, kinases, and hydrolases. Most of the hydrolases are proteases. There are also many proteins that are members of proto-oncogene families, as well as families of 'select regulatory molecules':

i) proteins involved in specific steps of signal transduction such as heterotrimeric GTP-binding proteins and cell cycle regulators, and
ii) proteins that modulate the activity of kinases, G proteins, and phosphatases.

Conservation of Core Processes

Reasonable comparative information is available for beginning the analysis of the evolution of the human genome. The genomes of *S. cerevisiae* ('bakers' yeast') and two diverse invertebrates, *C. Elegans* (a nematode worm) and *D. melanogaster* (fly), as well as the first plant genome, *A. thaliana*, recently completed, provide a diverse background for genome comparisons.

The concept of orthology is important because if two genes are orthologs, they can be traced by descent to the common ancestor of the two organisms and therefore are likely to perform similar conserved functions in the different organisms. It is critical in this analysis to separate orthologs (a gene that appears in two organisms by descent from a common ancestor) from paralogs (a gene that appears in more than one copy in a given organism by a duplication event) because paralogs may subsequently diverge in function.

Following the yeast-worm ortholog comparison in, HGP identified two different cases for each pairwise comparison (human-fly and human-worm).

The first case was a pair of genes, one from each organism, for which there was no other close homolog in either organism. These are straightforwardly identified as orthologous, because there are no additional members of the families that complicate separating orthologs from paralogs.

The second case is a family of genes with more than one member in either or both of the organisms being compared. Chervitz *et al*. deal with this case by analysing a phylogenetic tree that described the relationships between all of the sequences in both organisms, and then looked for pairs of genes that were nearest neighbours in the tree. If the nearest-neighbour pairs were from different organisms, those genes were presumed to be orthologs.

These nearest neighbours can often be confidently identified from pairwise sequence comparison without having to examine a phylogenetic tree. If the nearest neighbours are not from different organisms, there has been a paralogous expansion in one or both organisms after the speciation event. When this one-to-one correspondence is lost, defining an ortholog becomes ambiguous.

For the initial computational overview of the predicted human protein set, HGP could not answer this question for every predicted protein. Therefore, HGP consider only strict orthologs, i.e., the proteins with unambiguous one-to-one relationships. By these criteria, there are 2, 758 strict human-fly orthologs, 2031 human-worm. HGP define the evolutionarily conserved set as those 1, 523 human proteins that have strict orthologs in both D. *melanogaster* and C. *Elegans*.

The set of conserved proteins is not distributed among molecular functions in the same way as the whole human protein set. Compared with the whole human set, there are several categories that are over-represented in the conserved set by a factor of ~2 or more. The first category is nucleic acid enzymes, primarily the transcriptional machinery.

The basic transcriptional and translational machinery is well known to have been conserved over evolution, from bacteria through to the most complex eukaryotes. Many ribonucleoproteins involved in RNA splicing also appear to be conserved among the animals. Other enzyme types are also over-represented. Many of these enzymes are involved in intermediary metabolism. The only exception is the hydrolase category, which is not significantly over-represented in the shared protein set.

Proteases form the largest part of this category, and several large protease families have expanded in each of these three organisms after their divergence. The category of select regulatory molecules is also over-represented in the conserved set. The major conserved families are small guanosine triphosphatases (GTPases) and cell cycle regulators.

The last two significantly over-represented categories are protein transport and trafficking, and chaperones. The most conserved groups in these categories are proteins involved in coated vesicle-mediated transport, and chaperones involved in protein folding and heat-shock response. These observations provide only a conservative estimate of the protein families in the context of specific cellular processes that were likely derived from the last common ancestor of the human, fly, and worm.

HGP has compared the human genome with the other sequenced eukaryotic genomes at three levels to explore the molecular building blocks of the vertebrate taxon. They are: molecular functions, protein families, and protein domains.

Molecular differences can be correlated with phenotypic differences to begin to reveal the developmental and cellular processes that are unique to the vertebrates.

The most prominent human expansions are in proteins involved in the following factors:

i) acquired immune functions;
ii) neural development, structure, and functions;
iii) intercellular and intracellular signalling pathways in development and homeostasis;
iv) hemostasis; and
v) apoptosis.

The most striking differences between the human genome and the *Drosophila* or *C. elegans* genome is the appearance of genes involved in acquired immunity. This is expected, because the acquired immune response is a defence system that only occurs in vertebrates.

HGP observe 22 class 1 and 22 class 11 major histocompatibility complex (MHC) antigen genes and 114 other immunoglobulin genes in the human genome. In addition, there are 59 genes in the cognate immunoglobulin receptor family. At the domain level, this is exemplified by an expansion and recruitment of the ancient immunoglobulin fold to constitute molecules such as MHC, and of the integrin fold to form several of the cell adhesion molecules that mediate interactions between immune effector cells and the extracellular matrix.

Vertebrate-specific proteins include the paracrine immune regulators family of secreted 4-alpha helical bundle proteins, namely the cytokines and chemokines. Some of the cytoplasmic signal transduction components associated with cytokine receptor signal transduction are also features that are poorly represented in the fly and worm. These include protein domains found in the signal transducer and activator of transcription (STATs), the suppressors of cytokine signalling (SOCS), and protein inhibitors of activated STATs (PIAS).

In the human genome, there is a marked increase in the number of members of protein families that are involved in neural development. Examples include neurotrophic factors such as ependymin, nerve growth factor, and signalling molecules such as semaphorins, as well as the number of proteins involved directly in neural structure and function such as myelin proteins, voltage-gated ion channels, and synaptic proteins such as synaptotagmin. These observations correlate well with the known phenotypic differences between the nervous systems of these taxa, notably:

i) the increase in the number and connectivity of neurons;
ii) the increase in number of distinct neural cell types;
iii) the increased length of individual axons; and
iv) the significant increase in glial cell number.

A number of prominent protein expansions are involved in the processes of neural development. Of the extracellular domains that mediate cell adhesion, the connexin domain-containing proteins exist only in humans. These proteins, which are not present in the *Drosophila* or C. *elegans* genomes, appear to provide the constitutive subunits of intercellular channels and the structural basis for electrical coupling.

Pathway finding by axons and neuronal network formation is mediated through a subset of ephrins and their cognate receptor tyrosine kinases that act as positional labels to establish topographical projections. The probable biological role for the semaphorins (22 in human compared with six in the fly and two in the worm) and their receptors (neuropilins and plexins) is that of axonal guidance molecules. Signalling molecules such as neurotrophic factors and some cytokines have been shown to regulate neuronal cell survival, proliferation, and axon guidance. Notch receptors and ligands play important roles in glial cell fate determination and gliogenesis.

Human expanded gene families play key roles directly in neural structure and function. One example is synaptotagmin, originally found to regulate synaptic transmission by serving as a Ca^{2+} sensor during synaptic vesicle fusion and release. Of interest is the increased co-occurrence in humans of PDZ and the SH3 domains in neuronal-specific adaptor molecules; examples include proteins that likely modulate channel activity at synaptic junctions.

Several ion-channel families, including the EAG subfamily, the voltage-gated calcium/sodium channel family, the inward-rectifier potassium channel family, and the voltage-gated potassium channel, alpha subunit family. Voltage-gated sodium and potassium channels are involved in the generation of action potentials in neutons. Together with voltage-gated calcium channels, they also play a key role in coupling action potentials to neurotransmitter release, in the development of neurites, and in short-term memory. The recent observation of a calcium-regulated association between sodium channels and synaptotagmin may have consequences for the establishment and regulation of neuronal excitability.

Major classes of protein components in both the central and peripheral nervous system of vertebrates are myelin basic protein and myelin-associated glycoprotein. Myelin P0 is a major component of peripheral myelin, and myelin proteolipid and myelin oligodendrocyte glycopotein are found in the central nervous system. Mutations in any of these myelin proteins result mi severe demyelination, which is a pathological condition in which the myelin is lost and the nerve conduction is severely impaired. Humans have at least 10 genes belonging to four different families involved in myelin production, and possibly more-remotely related members of the MOG family.

Many protein families that have expanded in humans relative to the invertebrates are involved in signalling processes, particularly in response to development and differentiation. They include secreted hormones and growth factors, receptors, intracellular signalling molecules, and transcription factors.

Developmental signalling molecules that are enriched in the human genome include growth factors such as writ, transforming growth factor—β (TGF-β), fibroblast growth factor (FGF), nerve growth factor, platelet derived growth factor (PDGF), and ephrins. These growth factors affect tissue differentiation and a wide range of cellular processes involving actin-cytoskeletal and nuclear regulation. The corresponding receptors of these developmental ligands are also expanded in humans.

The interactions of several of these adhesion domains with extracellular matrix proteoglycans play a critical role in host defence, morphogenesis, and tissue repair. Consistent with the well-defined role of heparan sulphate proteoglycans in modulating these interactions, we observe an expansion of the heparin sulphate sulfotransferases in the human genome relative to worm and fly. These sulfotransferases modulate tissue differentiation.

A similar expansion in humans is noted in structural proteins that constitute the actin-cytoskeletal architecture. Compared with the fly and worm, we observe an explosive expansion of the nebulin, aggrecan, and plectin repeats in humans. These repeats are present in proteins involved in modulating the actin-cytoskeleton with predominant expression in neuronal, muscle, and vascular tissues. Comparison across the five sequenced eukaryotic organisms revealed several expanded protein families and domains involved in cytoplasmic signal transduction.

Signal transduction pathways playing roles in developmental regulation and acquired immunity were substantially enriched. There is a factor of two or greater expansion in humans in the Ras superfamily GTPases and the GTPase activator and GTP exchange factors associated with them. Although there are about the same number of tyrosine kinases in the human and *C. elegans* gencimes, in humans there is an increase in the SH2, PTB, and ITAM domains involved in phosphotyrosine signal transduction.

There is a twofold expansion of phosphodiesterases in the human genome compared with either the worm or fly genomes. The downstream effectors of the intracellular signalling molecules include the transcription factors that transduce developmental fates. Significant expansions are noted in the ligand-binding nuclear hormone receptor class of transcription factors compared with the fly genome, although not to the extent observed in the worm.

Perhaps the most striking expansion in humans is in the C_2H_2 zinc finger transcription factors. Pfam detects a total of 4500 C_2H_2 zinc finger domains in 564 human proteins, compared with 771 in 234 fly proteins. This means that there has been a dramatic expansion not only in the number of C_2H_2 transcription factors, but also in the number of these DNA-binding motifs per transcription factor.

Furthermore, many of these transcription factors contain either the KRAB or SCAN domains, which are not found in the fly or worm genomes. These domains are involved in the oligomerisation of transcription factors and increase the combinatorial partnering of these factors. In general, most of the transcription factor domains are shared between the three animal genomes, but the reassortment of these domains results in organism-specific transcription factor families.

The domain combinations found in the human, fly, and worm include the BTB with C_2H_2 in the fly and humans, and homeodomains alone or in combination with POU and LIM domains in a of the animal genomes. In plants, however, a different set of tanscription factors are expanded, namely, the myb family, and a unique set that includes VP1 and AP2 domain-containing proteins. The yeast genome has a paucity of transcription factors compared with the multicellular eukaryotes, and its repertoire is limited to the expansion of the yeast-specific C_6 transcription factor family involved in metabolic regulation.

Most of the protein domains are highly conserved. An interesting observation is that worms and humans have approximately the same number of both tyrosine kinases and serine/threonine kinases. These are merely counts of the catalytic domain, the proteins that contain these domains also display a wide repertoire of interaction domains with significant combinatorial diversity.

Hemostasis is regulated primarily by plasma proteases of the coagulation pathway and by the interactions that occur between the vascular endothelium, and platelets. Consistent with known anatomical and physiological differences between vertebrates and invertebrates, extracellular adhesion domains that constitute proteins integral to hemostasis are expanded in the human relative to the fly and worm. We note the evolution of domains such as FIMAC, FN1, FN2, and Clq that mediate surface interactions between hematopoeitic cells and the vascular matrix. In addition, there has been extensive recruitment of more-ancient animal-specific domains such as VWA, VWC, VWD, kringle, and FN3 into multidomain proteins that are involved in hemostatic regulation.

Although we do not find a large expansion m the total number of serine proteases, this enzymatic domain has been specifically recruited into several of these multidomain proteins for proteolytic regulation in the vascular compartment. These are represented in plasma proteins that belong to the kinin and complement pathways. There is a cant expansion in two families or matrix metalloproteases: ADAMA disintegrin and metalloprotease) and MMPs (matrix metalloproteases).

Proteolysis of extracellular matrix (EEM) proteins is critical for tissue development and for tissue degradation in diseases such as cancer, arthritis, alzheimer's disease, and a variety of inflammatory conditions. ADAMs are a family of integral membrane proteins with a pivotal role in fibrinogenolysis and modulating interactions between hematopoietic components and the vascular matrix components. These proteins have been shown to cleave matrix proteins, and even signalling molecules: ADAM-17 converts tumour necrosis factor-α, and ADAM-10 has been implicated in the Notch signa'ling pathway.

Evolutionary conservation of some of the apoptotic pathway components across eukarya is consistent with its central role in developmental regulation and as a response to pathogens and stress signals. The signal transduction pathways involved in programmed cell death, or apoptosis, are mediated by interactions between well-characterised domains that include extracellular domains, adaptor domains, and those found in effector and regulatory enzymes.

HGP enumerated the protein counts of central adaptor and effector enzyme domains that are found only in the apoptotic pathways to provide an estimate of divergence across eukarya and relative expansion in the human genome when compared with the fly and worm. Adaptor domains found in proteins restricted only to apoptotic regulation such as the DED domains are vertebrate-specific, whereas, others like BIR, CARD, and Bcl2 are represented in the fly and worm. Although plants and yeast lack the caspases, caspase-like molecules, namely the para- and meta-caspases, have been reported in these organisms. Compared with other animal genomes, the human genome shows an expansion in the adaptor and effector domain-containing proteins involved in apoptosis, as well as in the proteases involved in the cascade such as the caspase and calpain families.

There are fewer cytochrome P450 genes in humans than in either the fly or worm. Lipoxygenases, on the other hand, appear to be specific to the vertebrates and plants, whereas the lip-oxygenase-activating proteins may be vertebrate-specific. Lipoxygenases are involved in arachidonic acid metabolism, and they and their activators have been implicated in diverse human pathology ranging from allergic responses to cancers.

One of the most surprising human expansions, however, is in the number of glyceraldehyde-3-phosphate dehydrogenase (GAPDH) genes (46 in humans, three in the fly, and four in the worm). There is, however, evidence for many retrotransposed GAPDH pseudogenes, which may account for this apparent expansion. However, it is interesting that GAPDH, long known as a conserved enzyme involved in basic metabolism found across all phyla from bacteria to humans, has recently been shown to have other functions. It has a second catalytic activity, as a uracil DNA glycosylase and functions as a cell cycle regulator and has even been implicated in apoptosis.

Another striking set of human expansions has occurred in certain families involved in the translational machinery. HGP identified 28 different ribosomal subunits that each have at least 10 copies in the genome; on average, for all ribosomal proteins there is about an 8- to 10-fold expansion in the number of genes relative to either the worm or fly.

Retrotransposed pseudogenes may account for many of these expansions. A number of ribosomal proteins have secondary functions independent of their involvement in protein biosynthesis. Many of these expansions likely represent intronless paralogs that have presumably arisen from retrotransposition, and again there is evidence that many of these may be pseudogenes. A second form of this factor has been identified with tissue-specific expression in skeletal muscle and a complementary expression pattern to the ubiquitously expressed eEFIA.

Alternative splicing results in multiple transcripts from a single gene, and can therefore generate additional diversity in an organism's protein complement. HGP has identified 269 genes for ribonucleoproteins. This represents over 2.5 times the number of ribonucleoprotein genes in the worm, two times that of the fly, and about the same as the 265 identified in the *Arabidopsis* genome. Whether the diversity of ribonucleoprotein genes in humans contributes to gene regulation at either the splicing or translational level is unknown.

In post-translational modifications, the most prominent expansion is the transglutaminases, calcium-dependent enzymes that catalyse the cross-linking of proteins in cellular processes such as hemostasis and apoptosis. The vitamin K-dependent gamma carboxylase gene product acts on the GLA domain found in coagulation factors, osteocalcin, and matrix GLA protein. Tyrosylprotein sulfotransferases participate in the post-translational modification of proteins involved in inflammation and hemostasis, including coagulation factors and chemokine receptors.

Although there is no significant numerical increase in the counts for domains involved in nuclear protein modification, there are a number of domain arrangements in the predicted human proteins that are not found in the other currently sequenced genomes. These include the tandem association of two histone deacetylase domains in HD6 with a ubiquitin finger domain, a feature lacking in the fly genome. An additional example is the co-occurrence of important nuclear regulatory enzyme PARP (poly-ADP-ribosyl transferase) domain fused to protein-interaction domains—BRCT and VWA in humans.

There are several possible explanations for the differences in phenotypic complexity observed in humans when compared to the fly and worm. Some of these relate to the prominent differences in the immune system, hemostasis, neuronal, vascular, and cytoskeletal complexity. The finding that the human genome contains fewer genes than previously predicted might be compensated for by combinatorial diversity generated at the levels of protein architecture, transcriptional and translational control, post-translational modification of proteins, or post-transcriptional regulation.

Extensive domain shuffling to increase or alter combinatorial diversity can provide an exponential increase in the ability to mediate protein. Evolution of apparently new protein domains and increasing regulatory complexity by domain accretion both quantitatively and qualitatively are two features that we observe in humans. Perhaps the best illustration of this trend is the C_2H_2 zinc finger-containing transcription factors, where we see expansion in the number of domains per protein, together with vertebrate-specific domains such as KRAB and SCAN. Recent reports on the prominent use of internal ribosomal entry sites in the human genome to regulate translation of specific classes of proteins suggests that this is an area that needs further research to identify the full extent of this process in the human genome. At the post-translational level, although we provide examples of expansions of some protein families involved in these modifications, further experimental evidence is required to evaluate whether this is correlated with increased complexity in protein processing. Post-transcriptional processing and the extent of isoform generation in the human remain to be catalogued in their entirety.

The whole-genome sequencing approach versus BAC by BAC experience in applying the whole-genome shotgun sequencing approach to a diverse group of organisms with a wide range of genome sizes and repeat content allows us to assess its strengths and weaknesses.

With the assess of the method for a large number of microbial genomes, *Drosophila,* and now the human, there can be no doubt concerning the utility of this method. The large number of microbial genomes that have been sequenced by this method demonstrate that megabase-sized genomes can be sequenced efficiently without any input other that the *de novo* mate-paired sequences. With more complex genomes like those of *Drosophila* or human, map information, in the form of well-ordered markers, has been critical for long-range ordering of scaffolds.

For joining scaffolds into chromosomes, the quality of the map is more important than the number of markers *per se.* Although this mapping could have been performed concurrently with sequencing, the prior existence of mapping data was beneficial. During the sequencing of the *A. thaliana* genome, sequencing of individual BAC clones permitted extension of the sequence well into centromeric regions and allowed high-quality resolution of complex repeat regions.

Likewise, in *Drosophila,* the BAC physical map was most useful in regions near the highly repetitive centromeres and telomeres. WGA has been found to deliver excellent-quality reconstructions of the unique regions of the genome. As the genome size, and more importantly the repetitive content, increases, the WGA approach delivers less of the repetitive sequence.

The cost and overall efficiency of clone-byclone approaches makes them difficult to justify as a stand-alone strategy for future large-scale genome-sequencing projects. Specific applications of BAC-based or other clone mapping and sequencing strategies to resolve ambiguities in sequence assembly that cannot be efficiently resolved with computational approaches alone are clearly worth exploring. Hybrid approaches to whole-genome sequencing will only work if there is sufficient coverage in both the whole-genome shotgun phase and the BAC clone sequencing phase.

Low Gene Number in Humans

HGP has sequenced and assembled ~95 per cent of the euchromatic sequence of *H. sapiens* and used a new automated gene prediction method to produce a preliminary catalogue of the human genes. HGP found far fewer genes (26,000 to 38,000) than the earlier molecular predictions (50,000 to over 140,000). Whatever the reasons for this current disparity, only detailed annotation, comparative genomics, and careful molecular dissection of complex phenotypes will clarify this critical issue of the basic 'parts list' of our genome.

Certainly, the analysis is still incomplete and considerable refinement will occur in the years to come as the precise structure of each transcription unit is evaluated. A good place to start is to determine why the gene estimates derived from EST data are so discordant with HGP's predictions. It is likely that the following contribute to an inflated gene number derived from ESTs: the variable lengths of 3′ and 5′-untranslated leaders and trailers; the little-understood vagaries of RNA processing that often leave intronic regions in an unspliced condition; the finding that nearly 40 per cent of human genes are alternatively spliced; and finally, the unsolved technical problems in EST library construction where contamination from heterogeneous nuclear RNA and genomic DNA are not uncommon.

There are genes that remain unpredicted owing to the absence of EST or protein data to support them, although the use of mouse genome data for predicting genes should limit this number. As was true at the beginning of genome sequencing, ultimately it will be necessary to measure m-RNA in specific cell types to demonstrate the presence of a gene.

A population of organisms might have to pay a price for the number of genes it can possibly carry. When the number of genes becomes too large, each zygote carries so many new deleterious mutations that the population simply cannot maintain itself. The mammalian genome would contain a maximum of not much more than 30,000 genes. An estimate of 30,000 gene loci for humans was also arrived at by Crow and Kimura.

Muller's estimate for D. *melanogaster* was 10,000 genes, compared to 13,000 derived by annotation of the fly genome. These arguments for the theoretical maximum gene number were based on simplified ideas of genetic load that all genes have a certain low rate of mutation to a deleterious state. However, it is clear that many mouse, fly, worm, and yeast knockout mutations lead to almost no discernible phenotypic perturbations. The modest number of human genes means that we must look elsewhere for the mechanisms that generate the complexities inherent in human development and the sophisticated signalling systems that maintain homeostasis.

There are a large number of ways in which the functions of individual genes and gene products are regulated. The degree of 'openness' of chromatin structure and hence transcriptional activity is regulated by protein complexes that involve histone and DNA enzymatic modifications. The location, timing, and quantity of transcription are intimately linked to nuclear signal transduction events as well as by the tissue-specific expression of many of these proteins. Equally important are regulatory DNA elements that include insulators, repeats, and endogenous viruses; methylation of CpG islands in imprinting; and promoter-enhancer and intronic regions that modulate transcription.

The spliceosomal machinery consists of multisubunit proteins as well as structural and catalytic RNA elements that regulate transcript structure through alternative start and termination sites and splicing. Hence, there is a need to study different classes of RNA molecules such as small nucleolar RNAs, antisense riboregulator RNA, RNA involved in X-dosage compensation, and other structural RNAs to appreciate their precise role in regulating gene expression. The phenomenon of RNA editing in which coding changes occur directly at the level of mRNA is of clinical and biological relevance.

Examples of translational control include internal ribosomal entry sites that are found in proteins involved in cell cycle regulation and apoptosis. At the protein level, minor alterations in the nature of protein-protein interactions, protein modifications, and localisation can have dramatic effects on cellular physiology. This dynamic system therefore has many ways to modulate activity, which suggests that definition of complex systems by analysis of single genes is unlikely to be entirely successful.

The human genome is asymmetrically populated with G + C content, CpG islands, and genes. However, the genes are not distributed quite as unequally as had been predicted. The most G + C-rich fraction of the genome, H3 isochores, constitute more of the genome than previously thought and are the most gene-dense fraction, but contain only 25 per cent of the genes, rather than the predicted ~40 per cent. The low G + CL isochores make up 65 per cent of the genome, and 48 per cent of the genes. This inhomogeneity, the net result of millions of years of mammalian gene duplication, has been described as the 'desertification' of the vertebrate genome. Why are there clustered regions of high and low gene density, and are these accidents of history or driven by selection and evolution? If these deserts are dispensable, it ought to be possible to find mammalian genomes that are far smaller in size than the human genome. Indeed, many species of bats have genome sizes that are much smaller than that of humans; for example, *Miniopterus,* a species of Italian bat, has a genome size that is only 50 per cent that of humans.

DNA Sequence Variation

Although HGP has identified and mapped more than three million SNPs, this by no means implies that the task of finding and cataloguing SNPs is complete. These represent only a fraction of the SNPs present in the human population as a whole. Nevertheless, this first glimpse at genome-wide variation has revealed strong inhomogeneities in the distribution of SNPs across the genome. Polymorphism in DNA carries with it a snapshot of the past operation of population genetic forces, including mutation, migration, selection, and genetic drift. The availability of a dense array of SNPs will allow questions related to each of these factors to be addressed on a genome-wide basis.

SNP studies can establish the ranize of haplo types present in subjects of different ethnogeoorigins, providing insights into population and migration patterns. Although such studies have suggested that modem human lineages derive from Africa, many important questions regarding human origins remain unanswered, and more analyses using detailed SNP maps will be needed to settle these controversies. In addition to providing evidence for population expansions, migration, and admixture, SNPs can serve as markers for the extent of evolutionary constraint acting on particular genes. The correlation between patterns of intraspecies and interspecies genetic variation may prove to be especially informative to identify sites of reduced genetic diversity that may mark loci where sequence variations are not tolerated.

Remarkable heterogeneity in SNP density implies that there are a variety of forces acting on polymorphism-sparse regions may have lower SNP density because the mutation rate is lower, because most of those regions have a lower fraction of mutations that are tolerated, or because recent strong selection in favour of a newly arisen allele swept the linked variation out of the population. The effect of random genetic drift also varies widely across the genome.

The non-recombining portion of the Y chromosome faces the strongest pressure from random drift because there are roughly one-quarter as many Y chromosomes in the population as there are autosomal chromosomes, and the level of polymorphism on the Y is correspondingly less. Similarly, the X chromosome has a smaller effective population size than the autosomes, and its nucleotide diversity is also reduced. But even across a single autosome, the effective population size can vary because the density of deleterious mutations may vary.

Regions of high density of deleterious mutations will see a greater rate of elimination by selection, and the effective population size will be smaller. As a result, the density of even completely neutral SNPs will be lower in such regions. There is a large literature on the association between SNP density and local recombination rates in *Drosophila*, and it remains an important task to assess the strength of this association in the human genome, because of its impact on the design of local SNP densities for disease-association studies.

Genome Complexity

In organisms with complex nervous systems, neither gene number, neuron number, nor number of cell types correlates in any meaningful manner with even simplistic measures of structural or behavioural complexity. Nor would they be expected to this is the realm of non-linearities and epigenesis. The 520 million neurons of the common octopus exceed the neuronal number in the brain of a mouse by an order of magnitude.

It is apparent from a comparison of genomic data on the mouse and human, and from comparative mammalian neuroanatomy, that the morphological and behavioural diversity found in mammals is underpinned by a similar gene repertoire and similar neuroanatomies. For example, when one compares a pygmy marmoset to a chimpanzee, the brain volume of this minute primate is found to be only about 1.5 cm^3, two orders of magnitude less than that of a chimp and three orders less than that of humans.

Yet the neuroanatomies of all dime brains are strikingly similar, and the behavioural characteristics of the pygmy marmoset are little different from those of chimpanzees. Between humans and chimpanzees, the gene number, gene structures and functions, chromosomal and genomic organisations, and cell types and neuroanatomies are almost indistinguishable, yet the developmental modifications that predisposed human lineages to cortical expansion and development of the larynx, giving rise to language, culminated in a massive singularity that by even the simplest of criteria made humans more complex in a behavioural sense.

There are special cases of regulatory gene networks that have a disproportionate effect on the overall system. HGP has presented several examples of regulatory genes that are significantly increased in the human genome compared with the fly and worm. These include extracellular ligands and their cognate receptors as well as nuclear regulators, where a few proteins control broad developmental processes.

There are more than 30 different mathematical descriptions of complexity. However, we have yet to understand the mathematical dependency relating the number of genes with organism complexity. One pragmatic approach to the analysis of biological systems, which are composed of nonidentical elements, is through graph theory. The elements of the system can be represented by the vertices of complex topographies, with the edges representing the interactions between them.

Examination of large networks reveals that they can self-organise, but more important, they can be particularly robust. This robustness is not due to redundancy, but is a property of inhomogeneously wired networks. The error tolerance of such networks comes with a price; they are vulnerable to the selection or removal of a few nodes that contribute disproportionately to network stability.

Gene knockouts provide an illustration. Some knockouts may have minor effects, whereas others have catastrophic effects on the system. In the case of vimentin, a supposedly critical component of the cytoplasmic intermediate filament network of mammals, the knockout of the gene in mice reveals them to be reproductively normal, with no obvious phenotypic effects, and yet the usually conspicuous vimentin network is completely absent.

On the other hand, ~30 per cent of knockouts in *Drosophila* and mice correspond to critical nodes whose reduction in gene product, or total elimination, causes the network to crash most of the time, although even in some of these cases, phenotypic normalcy ensues, given the appropriate genetic background. Thus, there are no 'good' genes or 'bad' genes, but only networks that exist at various levels and at different connectivities, and at different states of sensitivity to perturbation.

Sophisticated mathematical analysis needs to be constantly evaluated against hard biological data sets that specifically address network dynamics. Nowhere is this more critical than in attempts to come to grips with complexity, particularly because deconvoluting and correcting complex networks that have undergone perturbation, and have resulted in human diseases, is the greatest significant challenge now facing us.

Complete sequencing of the human genome would open up new strategies for human biological research and would have a major impact on medicine, and through medicine and public health, on society. Effects on biomedical research are already being felt. This assembly of the human genome sequence is but a first, hesitant step on a long and exciting journey toward understanding the role of the genome in human biology. It has been possible only because of innovations in instrumentation and software that have allowed automation of almost every step of the process from DNA preparation to annotation.

We must define the complexity that ensues when this relatively modest set of about 30,000 genes is expressed. The sequence provides the framework upon which all the genetics, biochemistry, physiology, and ultimately phenotype depend. It provides the boundaries for scientific inquiry. The sequence is only the first level of understanding of the genome. All genes and their control elements must be identified; their functions, in concert as well as in isolation, defined; their sequence variation worldwide described; and the relation between genome variation and specific phenotypic characteristics determined.

Many diverse sources of data have shown that any two individuals are more than 99.9 per cent identical in sequence, which means that all the glorious differences among individuals in our species that can be attributed to genes falls in a mere 0. 1 per cent of the sequence. There are two fallacies to be avoided: determinism, the idea that ail characteristics of the person are hard-wired by the genome; and reductionism, the view that with complete knowledge of the human genome sequence, it is only a matter of time before our understanding of gene functions and interactions will provide a complete causal description of human variability.

REFERENCES

Bickmore, W. A., Sumner, A.T., *Trends Genet*, 5, 144, 1989.

Horvath, J.E., Schwartz, S., Eicher, E.E., *Genome Res*. 10, 839, 2000.

Nei, M., *Molecualr Evolutionary Genetics,* Columbia University Press, New York, 1987.

Warurton, P.E., Willard, H.F., in *Human Genome Evolution,* Jackson, M.S., Strachan, T., Dover, G., (eds.), BIOS Scientific, Oxford, 1996.

8

Genetic Applications

The application of genetic principles has played a central role in shaping human history and civilisation. Selection for desirable inherited traits led to the domestication of food plants and animals, which made agriculture possible. Since agriculture is a prerequisite for a civilisation to rise and flourish, it is no surprise that genetic improvement of agricultural species has continued from prehistory to the present day.

HUMAN GENOME PROJECT

Human Genome Project (HGP) is the coordinated effort of many laboratories to map the human genome, identify all of its genes, and eventually determine the entire genomic nucleotide sequence. Many different scientists are collaborating to complete the sequencing effort, and they acquire their sequences from the DNA of many different people.

Although DNA sequencing is a major part of the human genome project, the project also includes high-resolution mapping of human chromosomes using RFLPs and PCR-based genetic markers, cDNA cloning and sequencing, cloning large fragments with cosmid clones and YACs, mapping and sequencing genes in model organisms and comparing the results to human genes in order to speed the sequencing process, and managing the enormous amount of information generated by the project.

The ultimate goals of the project is to improve our understanding of the human genome and to apply that understanding for the benefit of humanity, primarily through medicine. In order to meet this goal, researchers had to develop the necessary technology for genetic mapping, DNA sequencing, and database management for the enormous quantity of data that would be generated. In spite of many obstacles and uncertainties, the technological advances were achieved, and in 1995 the project entered the final phase of massive DNA sequencing.

CHROMOSOME MAPPING

The main effort in the human genome project is high-resolution genetic mapping of the human chromosomes. Genetic map distances in centimorgans do not correspond in a linear fashion to numbers of nucleotide pairs, so the human genetic map is a product of the human genome project that stands separate from the DNA sequence but can be compared to it for understanding the relationship between genetic distances and physical distances in the genome.

The original goal was to obtain a map with at least one marker every two cm along each of the chromosomes by 1996, a goal that had already been achieved by that date. By 1990, researchers had mapped 6, 726 markers, 1, 867 of them representing known genes. By 1994, just four years later, researchers had mapped over 60, 000 markers. The potential benefits of mapping include identifying genetic markers that can be used for genetic testing and chromosome jumping and walking to locate genes.

Sequence Tagged Sites (STSs)

To develop DNA sequence of the human genome, researchers had to find a way to identify overlapping fragments of DNA.

Large DNA fragments have been cloned at random into YACs and cosmids. However, unless these fragments can be mapped in relation to one another, they are of little value. After a cloned fragment of DNA is assigned to a chromosome or a particular segment of a chromosome, it can be fingerprinted by mapping frequently repeated segments such as *Alu* elements, L1Hs elements, or poly (GT) repeats within the cloned fragment. These repeated segments form a particular pattern in a large clone.

If part of the pattern in one clone matches part of the pattern in another clone known to be on the same chromosome or within the same chromosomal segment, there is a strong possibility that the two clones overlap along part of their sequences. Overlapping clones are called contigs, because they represent contiguous segments. Contigs can be lined up with one another to form a linear map of overlapping contigs that represent large chromosomal regions. In some cases, entire chromosomes are now represented by a set of overlapping contigs. Once a set of contigs is identified, each contig can be sequenced to provide a set of overlapping DNA sequences that can be combined to provide the overall DNA sequence.

Using specific sites called sequence tagged sites (STSs), clones can also be ordered into contigs. STSs are short DNA segments that have been sequenced and assigned to a chromosomal location. Each STS can be amplified by PCR using defined primers. The sequence of each primer for STS amplification is available to researchers, so they can synthesise primers to amplify each STS.

Geneticists can test large clones, such as YAC or cosmid clones, for the presence of known STSs to determine their location and identify overlapping clones for identification of contigs. Geneticists can also identify contigs by aligning the overlapping DNA sequences of sequenced clones. After geneticists obtain the DNA sequence of a clone, they submit the sequence to a database, such as GenBank. DNA sequence databases contain the sequences of hundreds of thousands of cloned fragments of human DNA. Geneticists can take the sequence of any clone and use powerful computer programmes that are available on-line to compare the clone's sequence with all other sequences in the database.

Expressed Sequence Tags (ESTs)

Although the entire sequence of a chromosome is useful for many types of research, the most beneficial portions for medical applications are the protein-encoding sequences contained within mature mRNAs. Researchers can rapidly obtain such sequences by isolating all the mRNAs from different tissues, producing cDNA clones of the mRNAs, and then determining the sequences of the cDNA clones. DNA sequences obtained in this way are called expressed sequence tags (ESTs). A set of ESTs could be included as a large database of DNA sequences that researchers could access to match genes with an EST sequence. Such an EST database was initiated in 1991 but was compiled at a slow pace because an EST database was not originally envisioned as part of the human genome project.

Scientists who worked with ESTs found it difficult to obtain public funding. Recognising the commercial potential of EST databases, several private corporations initiated their own EST projects, in some cases maintaining rigid proprietary control over the DNA sequence information. A major financial barrier was overcome when the pharmaceutical corporation Merck and Co. committed to sponsoring the sequencing of 200,000 human ESTs, all for free public release.

A large EST database compiled from this project and a variety of other sources is now available *gratis* to all interested scientists worldwide via the internet. In many cases, known DNA sequences from other species have been used to identify the homologous human gene from the human EST database. For example, the cDNA sequence of the human telomerase gene was identified by comparing a yeast telomerase sequence with sequences in the human EST database.

HUMAN GENOME SEQUENCING

In haploid human genome about three billion nucleotide pairs are included. Human sequencing projects currently use automated DNA sequencers, which perform most of the sequencing steps automatically and provide the DNA sequence in computerised format. The sequences are entered into worldwide databases that researchers can access to compare a DNA sequence with others from different laboratories.

The GenBank database is one of several databases that exchange data on a daily basis to update sequences. With many laboratories contributing large amounts of information, the task of information management is potentially overwhelming. One of the major efforts in the human genome project has been the development of interactive computer databases and analysis programmes that researchers around the world can use in a coordinated effort. Such databases, as well as services for DNA sequence comparison, descriptions of genes, and integrated literature searching.

DNA sequencing is now much more rapid and inexpensive than it was just a few years ago. The human genome project is not the only beneficiary from improved sequencing methods. Scientists who conduct DNA analysis with any species can benefit from the improved efficiency of laboratory methods and information management. Major DNA sequence databases are not restricted to human DNA sequences but contain sequences submitted by researchers from around the world for many different species.

The broad benefits of improved technology and information management are among the strongest justifications for the large expenditures required by the human genome project. Biotechnology is now taking its place as one of the most significant industries of our time. The investments made in the human genome project will likely be repaid many times over as they are applied in biotechnology aimed at improvements in medicine, agriculture, industry and so on.

GENETIC TESTING

The major benefits of the human genome project is its contribution to the development of genetic tests. Inherited genetic disorders are often caused by recessive alleles that remain phenotypically hidden in heterozygotes. Other genetic disorders, such as Huntington disease, may be conferred by a dominant allele but do not appear until later in life because of delayed onset. Some genetic disorders, such as phenylketonuria, can be treated successfully if diagnosed early, before symptoms appear. Advances in genetic testing have made it possible to identify people who are heterozygous carriers of recessive alleles or people who are at risk for later onset of a genetic disorder.

Many genetic disorders can now be identified in a foetus during the early stages of pregnancy or even from a single cell extracted from a human embryo produced through *in vitro* fertilisation. Physicians now have at their disposal numerous genetic tests for identifying genetic disorders. These tests can provide physicians with information that they can use to develop effective prevention or treatment programmes. For example, when a physician determines from the results of a genetic test that a young woman has an inherited susceptibility for breast cancer, the physician can monitor that woman more closely for the early onset of breast cancer and can treat the cancer early, when the probability of successful treatment is high. Under such circumstances, genetic testing is clearly beneficial. However, as genetic testing for certain disorders has become widespread, serious legal and ethical questions have arisen regarding the proper use of genetic testing and the information derived from it.

Biochemical Genetic Tests

Genetic tests not based on DNA analysis have been available for many years. These tests are usually based on simple and inexpensive biochemical procedures of chromosomes. About 50 genetic disorders can be diagnosed with such tests, several using foetal tissue. Among the most widely administered genetic tests is newborn screening for phenylketonuria (PKU).

Every state in the United States as well as many foreign countries require mandatory screening of newborns for PKU. PKU is a devastating illness that causes severe mental retardation if left untreated. It is caused by the absence of functional phenylalanine hydroxylase due to homozygosity for mutant alleles at the PAH locus, which encodes the enzyme phenylalanine hydroxylase. PKU is present in about 1 in 12,500 newborns.

The Guthrie test requires a small amount of blood from an infant. The blood is placed on a filter paper and allowed to dry. The dried blood is then added to the growth medium of a bacterial strain of *Bacillus subtilis* that is auxotrophic for phenylalanine. High levels of phenylalanine in the blood cause rapid bacterial growth. When rapid growth is observed, the child's blood is then tested directly for phenylalanine and tyrosine levels to confirm the diagnosis. When evidence of PKU is present, physicians initiate extensive testing and treatment to prevent lifelong mental retardation in the affected children.

Sickle-cell anemia is due to homozygosity for the recessive *HBB*S* allele. Sickle-cell trait is much less severe than sickle-cell anemia and appears in people who are heterozygous for the *HBB*S* allele. The symptoms of sickle-cell trait are minor and do not have adverse consequences. However, one of the important symptoms of sickle-cell trait is increased susceptibility to streptococcal infection before 3 years of age, which currently causes unnecessary deaths among children with sickle-cell trait, particularly those who live in poverty and lack adequate medical care. Newborn screening and preventative treatment with antibiotics can avert potentially serious infections.

Genetic testing for sickle-cell trait does not have a sterling history. Since selection against malaria has increased the frequency of the *HBB*S* allele in certain regions of the world, sickle-cell trait is more common in certain ethnic groups, including those with ancestries originating in west Africa and the eastern Mediterranean. The frequency of heterozygous carriers of *HBB*S* among African Americans is quite high, about 1 in 12.

The test for sickle-cell trait was developed in the 1960s and was made available to the public as a voluntary test in some states. However, in some states the test was mandatory for African American children before they could enter school. Misunderstandings regarding the test, breaches of confidentiality for test results, and racial discrimination in its application have raised serious questions about how it should be used.

Aneuploidies can be detected by cytological examination of chromosomes prenatally in fetal cells and postnatally when symptoms in a newborn suggest the possibility of an aneuploid syndrome. Other currently incurable genetic disorders such as fragile-X syndrome and Huntington disease can be detected prenatally using recombinant DNA methods. Prenatal cytological tests are typically recommended by physicians for pregnant women who are 35 or older because of the increased probability of Down syndrome births. Although many genetic disorders cannot be cured, early detection assists in planning for appropriate education and medical care of the affected child.

DNA-based Genetic Tests

The human genome initiative has produced saturated genetic maps in which many genes responsible for genetic disorders have been mapped in relation to DNA marker loci. A large number of these genes have now been cloned and sequenced, and the DNA sequences of numerous mutant alleles have also been determined as part of the human genome project. The alleles responsible for over 200 genetic disorders can now be detected using DNA analysis.

DNA analysis can also identify people who are heterozygous for alleles; that confer recessive disorders and people who are prone to delayed onset of a genetic disorder. The first DNA tests for detection of mutant alleles were based on RFLP loci that were closely linked to the gene in question. However, the gene itself is a better marker than linked DNA marker loci for two reasons. First, because the gene itself is used, crossovers between the mutant allele and the DNA marker that is coupled to the allele are no longer a concern. Second, because several disorders are caused by multiple mutant alleles, a marker coupled to a mutant allele may only identify one of the several possible mutant alleles.

Using the cloned gene often makes it possible to identify many, if not all, mutant alleles responsible for a disorder. Huntington disease was among the first to be predicted using a linked RFLP locus. A specific RFLP marker was about 95 per cent accurate in predicting Huntington disease. The inaccuracies were due in part to rare crossovers between the RFLP marker locus and the HD locus. After the HD gene was cloned and sequenced, the RFLP test for Huntington disease became obsolete. The mutation that causes Huntington disease is a trinucleotide expansion that can be readily detected with a PCR test.

Recessive mutant alleles are the causing factors of most genetic disorders. Often, individuals with a recessive genetic disorder are compound heterozygotes for two different mutant alleles. This complicates genetic testing because each mutant allele may need to be identified with a separate DNA test, and certain mutant alleles may escape detection. In some disorders, such as fragile-X syndrome and Huntington disease, most mutant alleles are trinucleotide repeat expansions that can be readily detected with a simple PCR test. DNA testing for these genetic disorders is highly reliable. However, other genetic disorders, such as PKU and cystic fibrosis, may be caused by many different kinds of mutant alleles.

Tests for Mutant Alleles of Cystic Fibrosis Gene

Researchers can use sequence information to design highly reliable RFLP or PCR tests that directly identify specific mutant alleles. Among the most elusive yet important cases of this was cloning and sequencing the gene responsible for cystic fibrosis.

Cystic fibrosis is one of the most common genetic disorders in humans of European descent, with a frequency of mutant alleles of about q = 0.022 and a frequency of heterozygous carriers of about four per cent. Nucleotide sequence of the CF gene was reported in September 1989. After the gene was cloned and sequenced, information about cystic fibrosis came quickly. Researchers determined the amino acid sequence of the protein product from the DNA sequence. They identified the gene product as a large protein, that functions as a membrane transport protein for chloride ions.

When no functional protein is produced, chloride ion transport across cell membranes is reduced, which explains the abnormally high salt accumulation in the perspiration of people who are afflicted with the disorder. With the DNA sequence of the gene and the amino acid sequence of the gene product known, researchers could analyse the DNA sequences of the CF gene in people with cystic fibrosis to see what types of mutations caused the disorder.

About 70 per cent of people with cystic fibrosis have the same mutant allele, an allele with a deletion of three nucleotides (CTT) that encode phenylalanine at position 508 in the amino acid chain. Scientists have developed several DNA tests that detect this allele. The remaining 30 per cent of cystic fibrosis cases are due to over 100 different mutant alleles, which include substitution, deletion, frameshift, nonsense, and intron-splicing mutations. The type of mutation is related to the severity of symptoms in some cases. The wide variety of mutant alleles that cause cystic fibrosis significantly complicates the detection process. However, after a mutant allele has been identified in a person, and a specific test is designed to detect that allele, family members can be tested for that specific allele to identify heterozygous carriers.

DNA FINGERPRINTING

DNA fingerprinting is the identification of individuals based on a pattern of DNA markers the scientists detect in the genomic DNA of an individual. In most cases, DNA fingerprinting is based on minisatellite or microsatellite markers. The patterns detected in DNA fingerprinting are so variable among people that only identical twins share the same DNA fingerprints. In this section we examine how DNA fingerprints are used in forensic science and paternity testing, and how DNA fingerprinting has fared as evidence in court.

DNA Fingerprinting in Forensic Science

The technique of DNA fingerprinting was developed for the first time (1985, 86) by Alec Telly and his colleagues at Leicenster University in UK. In this field, establishment of the identity of a person, with the help of blood stains or hair roots will be possible, with almost absolute certainty. The term DNA finger print was coined to describe these unique patterns because they can be used in a manner familiar to that of a true finger print, to identify an individual.

Just as criminals tend to leave finger prints at the scene of crime, they may leave traces of their DNA behind in the form of skin, hair, blood or semen, which provide a source of DNA for forensic scientists. In this technique, DNA will be isolated from blood stains, semen stains or hair roots and will be subjected to southern blotting and DNA hybridisation, with the help of specific DNA probes. This will reveal polymorphism in DNA, which has a very stable inheritance. For this purpose, DNA from blood, semen or urine stains may also be amplified using PCR technique. The technique of DNA fingerprinting reveals such a great polymorphism that the possibility of two persons having same pattern of DNA fingerprint is very remote. Thus the DNA fingerprinting has tremendous power of exclusion. An innocent suspect can be cleared readily by a mismatch in DNA fingerprints. Exclusion is also possible on the basis of other types of evidence. Before DNA fingerprinting was available, exclusion of suspects was possible when blood types, or other types of biochemical markers, showed a mismatch. Mismatches in blood type or other markers can be used only to exclude suspects, not to prove identity. For example, if a culprit's blood found at a crime scene is type O, then a suspect with type O blood is not excluded but cannot be considered guilty on that basis because many people have type O blood.

DNA fingerprinting provides inclusive as well as exclusive evidence. For inclusive evidence, forensic scientists must generate DNA fingerprints using several different probes. When different probes are used to make several fingerprints, the likelihood that any two individuals chosen at random will have identical matches in all of them is extremely small.

In violent crimes, a criminal may remove the body of a victim from the crime scene, but remnants of blood or tissue from the victim may remain at the scene. DNA fingerprints from the crime scene can be matched to a body found elsewhere. Likewise, severely mutilated or decomposed bodies can be identified through DNA fingerprinting.

Although the DNA fingerprint of a missing person may not be available to match to a body, forensic scientists can use DNA fingerprints from a person's parents, siblings, or offspring to determine if blood or tissue from a crime scene or a recovered body matches a missing person. Such identification is possible because each DNA marker in a DNA fingerprint is inherited.

Fingerprinting in Paternity Testing

DNA fingerprinting is used widely for paternity testing and in criminal forensics. In most cases of questioned parentage, the mother is known, but the identity of the father is in question. A child inherits about half of the fragments in his or her DNA fingerprint from the mother and half from the father. Any fragments present in the child's DNA fingerprint that are not in the mother's DNA fingerprint must be inherited from the father.

Comparison of the child's DNA fingerprint with those of the mother and a suspected father can exclude or confirm the suspected father as the biological father. If the child has several fragments in his or her DNA fingerprint that are not present in the DNA fingerprint of either the mother or the suspected father, then the suspected father is not the child's genetic father. If, on the other hand, all fragments present in the child's DNA fingerprint that are not present in the mother's DNA fingerprint are present in the suspected father's DNA fingerprint, then the suspected father is the biological father.

Due to this tremendous power of both exclusion and inclusion, attorneys routinely use DNA fingerprinting as legal evidence. Laboratory techniques and assumptions of allele frequencies associated with DNA fingerprinting have been raised as reasons to exclude DNA fingerprinting as evidence. In recent years, however, the legal admissibility of DNA evidence for inclusion has gained widespread legal acceptance.

Problems of DNA Fingerprinting

DNA fingerprinting is now well established in developed countries. In Indian cintext, however, the awareness of the strength and potential of this technique is lacking. One of the major cause in their context is that this technique has not yet become the part of the Evidence Act. A powerful tool like DNA testing, if used properly by the expert, can help not only to find the culprit, but also can exclude a falsely accused. Individual from crime scene, probably a more meaningful outcome of this testing especially when primary aim of the judicial system is to find the truth.

CHROMOSOMAL ABERRATIONS

Since the development of chromosome technique in human beings, chromosome aberrations of patients in the hospitals are routinely analysed. As a result of such examinations, a number of abnormalities could be attributed to chromosomal aberrations.

Triploidy

A solitary case of triploid human being ($2n = 69$) was reported in 1960 by Brook and Santerson. This triploid individual had three complete sets of autosomes and XXY chromosomes. Consequently, it was a male individual and showed abnormalities in cerebral development, had syndactyly of hands and feet and was small jawed obese. Such triploid individual triploid individual survived only upto birth.

Aneuploidy and Structural Changes

Trisomics ($2n = 47$) and nonsomics ($2n = 45$) are known in human beings. Trisomy and monosomy may involve a sex chromosome or on autosome.

Aneuploidy Involving Sex Chromosomes

Aneuploid chromosome numbers involving X chromosome and resulting phenotypes are as follows:

Sex chromosomes	Sex	Phenotypes
XO (monosomic)	female	Turner's syndrome
XX (disomic)	female	normal
XXX (trisomic)	female	super female
XXXX (tetrasomic)	female	mental abnormalities

XXXXX (pentasomic)	female	mental abnormalities)
XY (disomic)	male	normal
XYY (trisomic)	male	normal
XXY (trisomic)	male	Klinefelter's syndrome
XXYY (tetrasomic)	male	
XXXY (tetrasomic)	male	Extreme Klinefelters
XXXXY (pentasomic)	male	

Turner's Syndrome

Turner's syndrome are characterised by monosomy of XO type. These are immature female (sterile) with webbed neck.

Klinefelter's Syndrome

Klinefelter's syndromes are characterised by trisomy (XXY). These are male individuals, who are phenotypically fairly normal but have a very law sperm count and are therefore steriles.

Mongolian Idiots or Down Syndromes

In trismoic-21, chromosome is a small chromosome and its addition does not cause lethality. Phenotypic abnormalities associated with these syndromes included slant of eyes, thick tongue, slagging mouth, unusual palm and feet, obesity and slow mental growth. The life expectancy of idiots is about 8 to 12 years.

GENETIC PHARMACOLOGY

The production of a pharmaceutical product in a host organism or cell culture using recombinant DNA is known as genetic pharmacology. Because the genetic code is the same in bacteria and humans, a bacterium can produce the same polypeptide as a human using the same gene, provided the gene has a bacterial promoter and is devoid of introns. Unfortunately, not all human proteins can be properly encoded and assembled in a bacterium.

Although bacterial and human cells use the same genetic code, they do not have the same apparatus for protein modification and processing. Although a bacterium can assemble the same linear array of amino acids as a human cell using the same gene, a bacterium cannot always produce a protein with the same secondary, tertiary, and quaternary structures as the native human protein.

A bacterial cell also cannot always attach proper additional compounds, such as phosphate groups, carbohydrates, and lipids, that a functional protein may need. These obstacles are formidable in some cases, but they are not always insurmountable. The case of recombinant human insulin, the first genetically engineered pharmaceutical to be licensed for human use, is a good example of how some innovative techniques can be used to overcome inherent problems and produce a genetically engineered product commercially.

Human Insulin Expression in Bacteria

Insulin is produced by b cells in the pancreas. It regulates sugar metabolism in the body. Insulin deficiency causes diabetes and often requires frequent hypodermic injections of insulin to maintain proper levels in the blood. Until recently, supplemental insulin was extracted from the pancreas tissue of cattle or pigs that had been killed for meat. Bovine (cattle) insulin differs from human insulin in three of the 51 amino acids, and porcine (pig) insulin differs by only one amino acid.

In spite of the similarities of bovine and porcine insulin to human insulin some patients mount an immune response to bovine or porcine insulin. In most cases, a switch to human insulin overcomes the problem. However, in a few cases, the immune response to animal insulin persists when the switch is made to human insulin. If a patient's treatments begin with human insulin, there is no adverse immune response.

Many people suffer from insulin-dependent diabetes, so there is a great need for human insulin. Human insulin is a protein that is needed commercially in abundance, so it was a prime candidate for genetic pharmacology. However, fully processed human insulin in the proper conformation could not be made in bacteria directly from a human insulin gene.

Insulin is encoded by a single gene that is transcribed and translated into a single polypeptide called preproinsulin. There are two segments in preproinsulin that are not present in the final form. The preamino acids are at the amino end of the polypeptide and facilitate secretion from the β cells. As the β cells secrete the preproinsulin, they cleave the pre- amino acids from the molecule to generate proinsulin.

The pro-amino acids are located in the centre of the polypeptide, in a segment called the C peptide, and help to form the proper tertiary and quaternary structures. After the proper structure is formed, the C peptide is cleaved from the molecule to form the final protein. The C peptide is present in the middle of the molecule, so its cleavage generates two polypeptides in the final insulin molecule with three disulfide bonds, two that connect the A and B chains to each other, and one that stabilises the tertiary structure of the A chain. Human insulin produced in bacteria requires mature A and B chains that can be connected with the appropriate disulfide bonds.

Researchers synthesised two artificial genes, one that encodes the A chain and the other that encodes the B chain. They attached the genes to the lacZ gene in the lac operon in such a way that the insulin codons were in frame with the lacZ codons, with an extra methionine codon placed between the lacZ gene and the insulin gene. This arrangement created fusion genes in which the insulin genes are under the control of the lac operon promoter and operator. The researchers placed the fusion genes into plasmids and transformed the recombinant plasmids into separate colonies of *E. coli*. When they grew the cells in the absence of lactose, the insulin genes were not expressed, and the cells grew and divided. When the bacterial cultures were transferred to a medium with IPTG (an inducer of the lac operon), the introduced genes were expressed.

The product of each introduced gene is a fusion protein with the left portion of the β-galactosidase protein (called a β-gal fragment) and either an A chain or a β chain of insulin attached to it by a methionine. The β-gal-Met-A and β-gal-Met-B fusion proteins are large and precipitate into inclusion bodies in the bacterial cells, where they are protected from bacterial enzyme destruction. If the small insulin chains were produced by themselves, bacterial enzymes would destroy them, so an attached polypeptide is needed.

Cyanogen bromide (CNBr) cleaves polypeptide chains on the carboxyterminal end of methionine only. Treatment of the isolated fusion proteins with cyanogen bromide cleaves the insulin chains from the β-gal fragments because of the methionine between them. Neither of the insulin chains has methionine in it, so CNBr leaves the insulin chains intact. After researchers have purified the A and B chains, they mix them and treat them chemically so that the disulfide bonds form.

The final product is genetically engineered human insulin. The method that is now the preferred choice for commercial production of recombinant human insulin does not require two separate bacterial cultures. Instead, an engineered sequence that encodes proinsulin is attached to the β-gal fragment. In this situation, the entire insulin molecule can be made in a single bacterial culture. The proinsulin is purified and separated from the β-gal fragment either by treatment with CNBr or through enzymatic means. Enzymes then remove the pro- amino acids to produce intact human insulin. Recombinant human growth hormone (HGH) was the second product of genetic pharmacology to be licensed for human use.

Foreign Protein Expression in Yeast

Genetic pharmacology is the production of medically useful proteins in bacterial, yeast, or mammalian cells. Several products produced by genetic pharmacology are currently being marketed. The enzymes bacteria uses for transcription and translation are different from those used in human cells. Although some eukaryotic proteins assume the correct structure when translated in a bacterial cell, there are other eukaryotic proteins that cannot be properly constructed in a bacterium.

Yeast species are single-celled eukaryotes that grow much like bacteria. Some proteins that fail to assume the proper conformation in bacteria can be produced successfully in yeast. The hepatitis B vaccine is among the first commercial successes with genetically engineered pharmaceuticals in yeast. The hepatitis B virus has a protein coat that certain human antibodies recognise as an antigen during a hepatitis B infection. Vaccines often contain weakened viruses or purified viral antigens that are then administered to the patient to elicit an immune response. However, in some cases, it is not possible to sufficiently weaken the virus or purify the antigen without contaminating the preparation with active viruses.

If active virus particles contaminate a vaccine, then the vaccine can cause the disease that it is intended to prevent. In the case of hepatitis B, a purified coat protein without the viral DNA would make an ideal vaccine. Researchers cloned the virus genes and chose the *HBsAg* gene because it encodes a protein that acts as an antigen on the virus's surface. They cloned the gene into a bacterial expression vector but could not use it to produce a functional antigen because of problems with protein structure. However, when they placed the gene into a yeast expression vector, the protein it encoded functioned as an antigen. Genetically engineered hepatitis B vaccine is now widely available.

Production of Human Proteins in Mammalian Cells

Some functional human proteins can be in bacteria with a little manipulation. All of these proteins are relatively simple in their structure. Certain complex proteins that cannot be produced effectively in either bacteria or yeast are needed in abundance for medical purposes. These proteins must be produced in mammalian cells that can process the proteins properly and fold them into their correct conformations.

Work is under way to produc. such proteins in genetically engineered dairy cattle so that the proteins are secreted in the milk. The cow's cells could process human proteins effectively and produce them in large quantities. Since the cattle themselves are genetically altered, they should pass the engineered gene to their progeny so that unlimited quantities of the protein can be produced. Even simple proteins now produced in genetically engineered bacteria and yeast could potentially be produced at far greater savings in the milk of dairy cattle. However, production of proteins in cattle is still in the experimental stages and has not yet been developed sufficiently to use commercially.

A readily available alternative is mammalian cell culture. Certain mammalian cells can be cultured, and express the products of foreign genes. Mammalian cells, however, are used only when bacteria and yeast systems do not work, because mammalian cells grow more slowly and are very expensive to maintain. Thus, the protein must be highly valuable for a mammalian cell system to be used. The first protein produced commercially in a mammalian cell system was tissue plasminogen activator, an enzyme that modifies plasminogen, a protein in the blood, into plasmin. Plasmin, in turn, digests another protein, fibrin, into smaller fragments. Fibrin is the protein that causes blood clots in coronary arteries that stop blood flow to the heart during a heart attack.

When plasminogen activator protein is administered to a heart attack victim, it changes plasminogen into plasmin, which then digests the clot and restores blood flow to the heart.

Researchers placed a plasminogen activator gene derived from cDNA under the control of a strong promoter and added the gene into a mammalian cell culture.

Attached to the gene is a second gene, called DHFR, that encodes the enzyme dihydrofolate reductase (DHFR). This enzyme provides resistance to the drug methotrexate, so methotrexate can be used to select cells with the plasminogen activator gene. As levels of methotrexate are increased, resistance only occurs when the DHFR gene is amplified into many copies. The cells that survive have multiple copies of the DHFR gene, which, in the process of amplification, multiplies the copy number of the closely linked plasminogen activator gene. The multiple copies of the plasminogen activator gene produce useful quantities of plasminogen activator protein.

HUMAN GENE THERAPY

Human gene therapy is focused on correcting the genetic cause of the disorder, which, in turn, alleviates the symptoms. Gene therapy can be divided into two types, somatic-cell gene therapy and germ-line gene therapy.

- *Somatic-cell gene therapy*: This type of gene therapy is based on the principle that a functional gene can be delivered to somatic cells in a tissue to correct a disorder in the cells. There are two types of somatic-cell gene therapy. The first is *in vivo* therapy, a procedure in which physicians introduce a functional gene into a virus and add the virus directly to the cells in the body so that the virus introduces the functional gene into the cells. The second type is *ex vivo* therapy, a procedure in which physicians extract somatic cells from a patient, insert a functional gene into the extracted cells, and return the cells to the tissue from which they were extracted.
- *Germ-line gene therapy*: Germ-line gene therapy is an alteration of the genetic composition of the germ line so

that people who carry a mutant allele no longer transmit that allele to their progeny. Technical and ethical constraints have prevented the development of human germ-line gene therapy. Since germ-line gene therapy has not been developed for use in humans, all references hereafter in this chapter to gene therapy refer to somatic-cell gene therapy.

Role of Viral Vectors in Gene Therapy

Ideally, the gene will be incorporated into the chromosomal DNA, where it will be expressed as if it were a part of the original genome. There are many ways to introduce DNA into cells, but the methods used most often require viral vectors. A retrovirus is a RNA virus that is reverse-transcribed into DNA after it has infected a cell. The DNA is then inserted into the chromosomal DNA of the host cell, where many copies of the virus can be produced by transcription.

Retroviruses can be engineered so that most of the viral genome is replaced by the functional human gene. These retroviral vectors can transfer an engineered gene into the chromosomes after the virus has infected the cell. Since the process utilises infection and gene transfer, it is called transfection. Once inside the cell, the engineered virus is reverse-transcribed into DNA, and the engineered gene is incorporated into the chromosomal DNA. Researchers construct viral vectors by deleting the viral genes that are necessary for replication, lysis, and infection of other cells. Consequently, the engineered virus cannot spread to cells other than those that are directly transfected.

Retroviruses are excellent choice as vectors for cancer gene therapy because the virus may target actively dividing tumour cells. This characteristic limits the use of retroviruses in other types of gene therapy because tissues that do not contain a large fraction of actively dividing cells cannot be successfully treated with retroviral vectors. Under these circumstances, adenovirus vectors are often used. Adenoviruses are double-stranded DNA viruses that insert themselves into the chromosomal DNA of the cells they infect.

Treatment of ADA Deficiency with Gene Therapy

The ADA locus on chromosome 20 encodes the enzyme adenosine deaminase (ADA), which degrades excess adenosine monophosphate (AMP) into uric acid. When the functional ADA is absent, the excess AMP is converted into compounds that are toxic to lymphoblasts, a type of cell that is active in the immune system. ADA deficiency causes severe combined immune deficiency, a disease in which the immune system fails to fight infections.

People with ADA deficiency may die from infections that are usually harmless to most people. ADA deficiency has traditionally been treated with transplants of bone marrow that produces functional ADA. ADA deficiency was a prime candidate for gene therapy for a number of reasons.

i) it required the addition of a single enzyme encoded by a single gene.
ii) it was known to be curable by introduction of foreign cells. Bone marrow transplants had been successful in the past, but their use was limited by the need to find an immunologically compatible donor.
iii) ADA deficient cells contain toxic levels of metabolites, which prevent them from growing and dividing.

Cells with a functional ADA gene that have been introduced into a patient should have a selective advantage over the ADA deficient cells. The first experiments were conducted on a 4-year-old ADA-deficient girl. Some of her T cells were isolated and incubated with interleukin-2 to stimulate growth.

Physicians inserted a functional ADA gene, together with a neomycin resistance marker gene, into the cells using a retroviral vector. They then reintroduced the engineered cells into the bone marrow and followed their progress by extracting blood samples and using PCR amplification of the neomycin marker as an indicator that engineered T cells were still present. The initial treatment has been followed by periodic maintenance infusions of engineered T cells. The treatment was successful, and the patient now lives a normal life. A second ADA-deficient child has received gene therapy with similar success.

CLINICAL GENETICS

Clinical genetics is defined as the diagnosis of genetic disorders and the care of individuals with genetic disorders. The phenotypes associated with most genetic disorders are characterised as syndromes, a collection of symptoms that affect different organs and tissues.

Clinical geneticists often work closely with physicians and practitioners in many other specialties to ensure that patients with a genetic disorder receive appropriate treatments for the variety of symptoms that accompany each disorder. Clinical genetics starts with the primary care physician, who must recognise the collection of symptoms that may be associated with a genetic disorder. Primary care physicians must therefore maintain a current understanding of advances in medical genetics.

Clinical geneticists review the symptoms along with the family history to determine what genetic disorders may be possible. The geneticist then determines which diagnostic tests should be utilised to accurately determine which, if any, known genetic disorder is the cause of the symptoms. Advances in cytogenetics, biochemical genetics, genetic mapping, and gene cloning and sequencing have greatly improved the accuracy and range of diagnostic genetic tests currently available for clinical use.

9
Issues in Genetics

Genetics is now more relevant to the public than ever before, underscoring the need for appropriate public education in science. As genetics has assumed an increasingly important role in medicine and agriculture, legal and ethical issues have become more common and more complicated. Genetic discoveries and applications may be protected legally. In most cases patent protection is sought.

EUGENICS

Concept of genetic purity in humans has existed throughout recorded history. Beliefs of racial or hereditary birthrights have been a part of nearly every human culture, and are evident in historical literature. Some cultures, political regimes, and religions have encouraged people to marry within their own ethnic or familial background in order to preserve perceived racial or ethnic purity, often with the notion that certain races or ethnic groups are superior and should not be contaminated by those that are inferior.

The idea of improving humans through selection in the same way that breeders improve animals and plants has been voiced for centuries. Sir Francis Galton, was among the most outspoken proponents of improving the human population by discouraging reproduction among those people that he considered to be the inferior members of society. He coined the term 'eugenics' in 1883 to describe the intentional hereditary improvement of the human population through selection.

During the early part of the twentieth century, up until the 1940s, eugenics was popular among philanthropists, scientists, politicians, lawyers, judges, journalists, and the better-educated and wealthier members of society. During this period, the idea that one of the most lofty goals of genetics is to improve the human race appeared in countless scientific papers and usually was the concluding subject of genetics textbooks.

Before long, some applications of eugenics made their way into law. Some people attributed traits such as imbecility, feeblemindedness, criminal behaviour, lack of intelligence, antisocial behaviour, insanity, idiocy, epilepsy, chronic alcoholism, and many others to heredity and proposed that society could rid itself of those traits by preventing people who had them from reproducing.

During 1896, some state legislatures passed laws forbidding the marriage of a woman and a man when both were considered imbeciles, feebleminded, or epileptic. Mandatory sterilisation laws began in 1907. Governments in United States and Europe passed laws requiring that criminals or people in mental institutions be sterilised to prevent transmission of their behavioural characteristics to subsequent generations.

In Europe, it was during the Nazi regime, eugenics reached its height. During the Holocaust, six million people were killed in mass executions because their genetic heritage was considered inferior. Such events in history are tragic reminders of how ideas about eugenics can be taken to extremes. The Nazi Holocaust is not that far removed from us in time; some of its survivors are alive today. Executions, incarceration, genocide, and rape have been justified during recent conflicts as a means of preserving ethnic superiority and racial purity.

Advocates of eugenics justify its practice on the assumption that preventing people with a genetic disorder from reproducing should eliminate, or significantly reduce, the frequency of an undesired trait in human populations.

However, many people consider the right to reproduce to be a basic human right, which is violated by involuntary sterilisation or laws that forbid certain people from reproducing. The United Nations' Universal Declaration of Human Rights, International Covenant of Political and Civil Rights, and the European Convention on Human Rights are just a few of the more prominent agreements on basic universal human rights that include the right to reproduce.

The effect of selection is usually negligible in altering the frequency of recessive genetic disorders. But laws that forbid marriages between close relatives can have an appreciable effect in preventing recessive genetic disorders. When inbreeding coefficients are high because of matings between close relatives, the frequency of recessive disorders increases considerably. Laws that prohibit marriages between close relatives are far more effective than mandatory sterilisation in preventing recessive genetic disorders. Most states prohibit marriages of first cousins and more closely related people for this reason.

There are no theoretical or empirical grounds for most eugenic practices. The effects of eugenic practices and modern technologies in altering allele frequencies through selection in humans are usually negligible. Like eugenic measures of the past, our current practices have little effect on overall allele frequencies. Unlike the eugenics of the past, the aim of modern practices is personal in nature, rather than an attempt to improve society. Some of the current practices include technologies that provide prenatal testing, donor screening in sperm banks, sex selection and so on.

Prenatal testing: It is the testing of a foetus to determine whether or not it has a genetic disorder and provides couples with information that permits them to choose an abortion rather than have a child with a genetic disorder. Voluntary abortion of fetuses with genetic disorders has decreased the frequencies of children born with certain genetic disorders. Abortion of foetuses that have a genetic disorder is a form of selection against the alleles that confer genetic disorders.

Donor screening in sperm banks: It is the selection of semen donors to exclude alleles that may cause genetic disorders. Many couples cannot have children because of male infertility. For a variety of reasons, a woman may undergo artificial insemination to conceive a child. Some hospitals and clinics maintain sperm banks in which they store semen samples provided by donors, usually medical students and physicians, who have been screened for family histories of genetic disorders. Sperm banks usually maintain anonymity for donors, although they keep genetic records on file in case questions arise later. With DNA tests now available, some hospitals and clinics screen donors to identify heterozygous carriers of recessive alleles that cause genetic disorders. Such screening is a mild form of selection that will have a negligible effect on the frequency of genetic disorders.

Sex selection: Sex selection is a practice of selectively biases and the probability of bearing a child with a particular sex is now available for humans. Animal breeders were the first to develop sex selection technologies. For example, some dairy operations use sex selection technologies to increase the frequency of female calves. Medical researchers have adapted the procedures for use in humans. Medical technicians can sort sperm cells by centrifugation (X-bearing sperm sediment at a different rate than Y-bearing sperm), by separation in a flow cytometer (a machine that runs sperm cells one by one through a capillary tube and directs X-bearing sperm into one container and Y-bearing sperm into another using an electrical field), or by allowing sperm to move through a column of albumin (in which X- and Y-bearing sperm move at different rates). None of the methods ensures that a child of a particular sex will be born. A few Y-bearing sperm may contaminate a sample of X-bearing sperm and vice versa, but the ratios are substantially biased toward the desired sperm-cell type. The prospective mother is artificially inseminated with the chosen sperm sample, and the probability of her bearing a child of the desired sex is greatly increased.

Currently, sex selection is expensive, and very few people use it. Consequently, its effect on the overall sex ratio is negligible. Should it become a widespread practice, concerns over alteration in the human sex ratio may be justified.

GENETIC SCREENING

Medical practitioners can treat certain genetic disorders if they diagnose the disorder early in the patient's life. Phenylketonuria (PKU) screening is an example of genetic screening that has gained widespread success and acceptance. Most states test all newborns for PKU to identify the disorder so that a physician can treat it before serious symptoms develop. The state usually funds PKU testing from tax revenues. There is a clear ethical obligation to treat a child with PKU and prevent mental retardation. There is also a financial benefit to doing so. People with mental retardation require expensive care for education and medical treatment. Often, the state bears much of the cost.

Expenditures that the state makes to detect PKU are repaid by prevention of mental retardation. For this reason, most states have enacted genetic screening laws when mandatory or voluntary screening can detect treatable genetic disorders. States have enacted laws for widespread screening of PKU, sickle-cell anemia, sickle-cell trait, and Tay-Sachs disease. In 1976, a federal law consolidated previous federal legislation into the US National Genetics Diseases Act, which provided federal funding for genetic screening, counselling, research, and education. US Congress rescinded the act in 1981 and diverted funding to the states.

Tay-Sachs disease is most common among people of Eastern European Jewish descent, and Jewish political groups have provided strong support of screening programmes. Screening for sickle-cell trait, on the other hand, has met much opposition because it has focused on certain ethnic groups and provided a basis for racial discrimination.

CHALLENGES OF GENETIC TESTING

Some genetic disorders that are not particularly difficult to deal with in daily life might be dangerous in certain occupational situations. The best-studied examples is glucose-6-phosphate dehydrogenase (G6PD) deficiency, caused by lower-than-normal levels of the enzyme glucose-6- phosphate dehydrogenase. When they are exposed to certain chemical agents, particularly oxidizing compounds, people with this disorder may have an increased tendency to suffer from anemia, although this conclusion has not been adequately demonstrated. Several recessive X-linked alleles at the G6PD locus cause G6PD deficiency, so it appears more often in men than in women.

Some mutant alleles provide resistance to malaria when heterozygous in women, so the frequencies of these alleles are higher in people with ancestry from tropical regions where malaria was common. Since some oxidizing chemicals may pose a more severe health threat to people with G6PD deficiency than to others, industrial employers may consider people with G6PD deficiency a greater health risk than those without the disorder. A simple and inexpensive biochemical test can detect G6PD deficiency, so a number of companies have screened workers for G6PD deficiency as part of the hiring process when the work may entail exposure to oxidizing agents.

Since G6PD deficiency is more common in men and among certain ethnic groups, this screening practice may be considered a type of gender, ethnic, or racial discrimination, and in this sense could be challenged legally. On the other hand, companies justify screening programmes by arguing that they are fulfilling their legal and moral obligation to prevent injuries and damage to worker health. In the 1970s, several companies in the United States conducted genetic screening for G6PD deficiency, sickle-cell trait, and a-antitrypsin deficiency among potential and current employees. In some cases, they used the results to deny employment or to alter worker responsibilities.

As the practice of genetic screening received adverse public attention, some companies terminated their screening programmes. However, a survey sponsored by the US Congress found that 10 per cent of responding companies conducted some form of genetic screening. Existing laws that regulate discrimination on the basis of ethnic origin and disabilities could conceivably restrict the extent to which employers can use occupational genetic testing. In particular, the 1990 Americans with Disabilities Act (ADA) forbids discrimination against workers with disabilities that do not seriously inhibit their performance on the job. To what extent this law applies to actual or potential genetic disorders will probably be decided in the courts.

Genetic risks identified by insurance companies include information gleaned from genetic testing as well as examination of medical and family histories. Most employers pay for a large proportion of the insurance coverage for their employees and thus have an added incentive to work with insurance companies to identify employees or employee dependents who are at risk for a genetic disorder. In some cases, they may exclude or reduce health and life insurance benefits on the basis of predisposition for expensive medical care or early death. With an increase in the number of genetic tests available, the potential for insurance companies to use genetic testing and analysis to determine health care risks is growing. In response, many scientists and physicians have recommended legislation to prevent insurance companies and employers from using genetic information to deny or alter health care coverage.

The Working Group on Ethical, Legal, and Social Implications of Human Genome Project and the US National Action Plan on Breast Cancer, with an endorsement from the US National Advisory Council for Human Genome Research, have proposed the following guidelines to prevent insurance providers from discriminating on the basis of genetic information:

i) Insurance providers should be prohibited from using genetic information or an individual's request for genetic services, to deny or limit any coverage or establish eligibility, continuation, enrollment, or contribution requirements.

ii) Insurance providers should be prohibited from establishing differential rates or premium payments based on genetic information or an individual's request for genetic services.

iii) Insurance providers should be prohibited from requesting or requiring collection or disclosure of genetic information.

iv) Insurance providers and other holders of genetic information should be prohibited from releasing genetic information without prior written authorisation of the individual. Written authorisation should be required for each disclosure and include to whom the disclosure would be made.

Challenges of DNA testing

DNA fingerprinting is a powerful forensic tool, which law enforcement officials can use to identify criminals and exonerate innocent suspects. Likewise, it is used routinely to identify paternity. In spite of its value, attorneys have challenged its accuracy and admissibility as evidence in court. Those who challenge the admissibility of DNA fingerprinting usually base their arguments on the possibility of faulty technique or potential inaccuracy of statistical assumptions.

There are several types of potential technical errors. The purity and identity of the DNA sample being tested must not be in question. Sources of DNA that belong to a criminal may come from sources at the scene of a crime, such as tissue under the fingernails of a victim or from semen in a rape case. Under such circumstances, the identity of the DNA source may be certain. The identity of blood at the scene of a crime, however, may be less certain. Also, the DNA must not be contaminated with DNA from another human. If it is, then DNA fingerprints of two or more people may be superimposed on one another. Any alterations in the methods used for electrophoresis, blotting, hybridisation, probe detection, and PCR can potentially cause artifacts that may lead to false conclusions.

Most laboratories that provide DNA testing services avoid technical error through the use of rigorous controls. They identify possible contamination by including DNA from the suspect, the victim, and laboratory workers in the analysis. They prevent technical problems by providing extensive training to technicians and by using high-quality equipment and chemicals. They often replicate experiments to ensure that the results are the same each time.

In recent years, DNA fingerprinting has established itself as a reliable method for identification. Courts usually determine the admissibility of scientific evidence by the degree of consensus among experts in the scientific community. If most experts in the field accept a technique or assumption as reliable, then it is admissible in court. Significant disagreement among experts about the reliability of a technique or assumption may be grounds for inadmissibility of evidence that is based on the technique. DNA fingerprinting has suffered both of these fates in court. Numerous convictions have been made when DNA fingerprinting was the principal evidence.

Informed consent and confidentiality

Clinical geneticists and physicians have relied on blood typing, biochemical tests, cytological tests, and pedigree analysis for decades. However, the number of DNA tests that are now available is numerous and grows significantly each year. With so many ways to test people for genetic disorders and identify people with DNA fingerprints, issues of informed consent and confidentiality are of great importance.

Informed consent for a genetic test is the voluntary submission to the test only after the person receives counselling about the test, what it will and will not identify, what the risks are, and how the information can be used once it is obtained.

Confidentiality is a person's right to be certain that those who obtain information will guard it from disclosure to unauthorised individuals or groups. Sharing information with potential employers, insurance companies, law enforcement officials, physicians, researchers, and other interested parties may violate a person's legal right to privacy. Confidentiality has often been breached in genetic testing. Results from tests identifying sickle-cell trait, XYY males, G6PD deficiency and several other genetic traits have not always been held in strict confidence, causing legal, social, and financial problems for the people identified with these traits.

Human geneticists who possess records on familial genetic disorders have sometimes denied other researchers access to the records in order to protect a family's right to privacy. In some cases, questions regarding access to records for the purpose of research are resolved only after a legal dispute. In such cases, the need for advancement in medical and scientific research must be weighed against the right to privacy and laws designed to protect that right.

Many legal questions regarding informed consent and confidentiality are yet to be resolved. As the number of genetic tests increases and as technical questions surrounding the reliability of these tests are raised, the legal questions may become even more difficult to answer. Informed consent and confidentiality are important issues for people who undergo any medical procedure.

INTELLECTUAL PROPERTY RIGHTS

Foundations and governments invest large sums of money in the development of medical and agricultural applications in genetics. Because genotypes and genes can often be propagated from a small number of individuals, private biotechnology companies and public institutions seek legal protection to prevent others from exploiting the products of their investments. Companies also invest much in developing techniques that are unique and profitable and seek protection of their investment. In many cases, the protection allows the company to require that others who use the techniques pay royalties for their use. The Constitution of the United States as well as numerous laws provide means for limited protection of inventions.

When someone clones a gene or discovers its sequence, genes can be patented. For example, the genetically engineered genes that produce human insulin in bacteria have been patented and have been the subject of a drawnout and expensive court battle. Even natural human genes can be patented. The long search for the gene responsible for cystic fibrosis (CF) culminated in a rush to patent the gene.

If an inventor discloses an invention in an article before patenting the invention, the invention becomes part of the public domain and cannot receive patent protection. In the case of the CF gene, the discoverers had submitted the articles for publication, but the articles were not scheduled to appear in print until after the patent applications had been filed. However, leaks to the press produced a flurry of questions about the gene, forcing an earlier-than-expected disclosure of the gene and its sequence.

Genetic traits themselves may be patented. For example, high-lysine maize derived from tissue culture and the use of cytoplasmic male sterility in seed production are genetic traits in plants that have been patented. Development of new techniques is important for advancement in research. Much effort and funding has been invested in the human genome project to find more methods of DNA sequencing. With large investments being made in development of techniques, patent protection of new techniques has been sought and awarded.

Protection rights of plants

The first plant genetic protection law was the US Plant Patent Act, passed in 1930. It extended plant patent protection for a period of 17 years, but only to asexually propagated varieties, such as fruit trees, grapes, and ornamental species. The protection did not extend to seed-propagated varieties.

Universities and companies produced seed-propagated plant varieties for decades with no legal protection, and farmers often sold seed without paying any royalties to the company that developed the variety. For a time, state-owned universities developed most pure-line varieties for self-pollinated species, such as wheat, barley, soybean, and oats.

Agricultural development of the state, rather than profit from the sale of seed, was the principal motive. Private companies, however, produced most F_1, hybrid varieties. Since F_2 seed loses its heterotic potential, a farmer gains nothing from saving seed. The genetic system of the plant ensures that a company's seed is protected, as long as the inbred lines used to produce F_1 seed are guarded.

However, when private companies began to market pure-line varieties and considered the genotype proprietary, the first attempt by US Congress to provide protection of pure-line varieties and inbred lines was the Plant Variety Protection (PVP) Act of 1970, which gave protection to the genotype of a pure-line variety or inbred line. A PVP certificate provides 18 years of protection to a variety or inbred line. During the time of protection, the PVP certificate owner has exclusive rights to propagate and market the variety or to use the inbred line in hybrid seed production.

There were two aspects of the PVP Act that some seed companies found objectionable. First, a farmer could purchase a protected variety then save some of the seed for planting in subsequent seasons without paying royalties to the company. Second, breeders could use the protected variety as a parent in hybridisation schemes to produce their own varieties without paying royalties to the PVP certificate owner when the new variety was marketed.

Due to these objections, seed companies sought utility patents on pure-line varieties in order to obtain protection for any type of commercial propagation and for the right to use the variety as a parent in the development of new commercial varieties. The US Patent Office denied such patents on the grounds that pure-line varieties were covered by the PVP Act. Under the Ex Parte Hibberd decision, the Board of Patent Appeals and Inferences reversed this policy and granted the right to patent plant varieties produced from seed, during 1985.

Some countries have passed legislation that prohibits transfer of their genetic resources outside of the country without compensation to the country. Unfortunately, many of these same countries fail to provide the financial backing necessary to adequately preserve the genetic resources they have legally restricted, causing irreplaceable loss of their protected resource. In addition, genetic resource protection laws are often not enforced, so genetic material may spread to other countries by contraband.

According to some legal experts, plant genetic resources be treated much like minerals or other tangible resources that are exported. The country of origin would receive royalties on new varieties developed from their older varieties. In many cases, it would be difficult to ascertain how much of a new variety's genotype was derived from the older variety and how much the royalties should be. The costs of litigation in such a situation may be so prohibitive that breeders may not wish to use protected resources to expand the genetic base of their varieties. Because the genetic foundations of most crop species are narrow and should be expanded, such a scenario is good.

According to many scientists the rights to use a plant variety as a parent in a breeding programme should be freely available to anyone in the world, but the commercial propagation and sale of a variety should be protected. In other words, a genotype can be protected, but its genes cannot. Seed companies usually oppose that policy because patents provide protection of a variety's use as a parent. In addition, some opponents of that policy believe that it discriminates against less developed countries because many of those countries rarely produce varieties that are used commercially. Such a policy could divert most benefits of protection to the economically developed countries. This problem became one of the most debated questions in the 1992 Earth Summit held in Rio de Janeiro.

Protection Rights of Animals

The genotypes of animals produced by traditional breeding cannot be protected, in part because each individual's genotype is different. However, an altered gene carried by an animal may be protected. Any animal carrying a gene altered by genetic engineering can be patented, and the patent extends to progeny that carry the gene. Also patented are inbred laboratory strains of mice that carry genes important to biomedical research. A 1987 ruling by the US Patent Office provided protection to animals that carry altered genes inserted into their genomes through genetic engineering.

The most important question raised by legal protection in genetics is whether legal protection promotes or inhibits the overall progress of research. On the one hand, those in favour of protection argue that products must be protected to allow investors to recover their investments. Without the promise of legal protection, many companies would not risk the investment necessary to develop new products and techniques, and progress in private research would be seriously jeopardised. On the other hand, legal restrictions on the use of a product or technique may inhibit beneficial research.

Scientists have suppressed or delayed publication of significant results in order to obtain patent rights. Products such as patented enzymes or test animals for use in research can carry such a high price that researchers may need to divert limited research funds from other uses in order to purchase the patented product. Patented procedures may carry restrictions or requirements that royalties be paid if the procedures are employed in experimental research with commercial applications, inhibiting the widespread use of such procedures. Striking a balance between the need to protect investment in research and the need to encourage cooperative research is not easy. Fortunately, some companies that own patents allow exemptions from some patent restrictions for non-profit research institutions, requiring royalties only when products that have been developed using a patented product are marketed.

Even so, the race to patent products and procedures in genetics research has accelerated. As the value of biotechnological products grows, so does litigation and the costs associated with it. Even universities and non-profit or governmental institutions have found patenting to be a way of raising needed funds. As the cost of research has risen at a time when public funding is difficult to obtain, universities have found legal protection through patents to be a means of raising funds to continue research. The university patents any product developed in its laboratories, then licenses the product to a private company, which pays royalties to the university for the right to market the product. In this way, the university can reap some of the financial benefits of the product without jeopardising its non-profit status by directly involving itself in marketing. Perhaps the greatest beneficiary of biotechnology patents is the legal profession. Law firms that specialise in biotechnology law are growing in number, and the need for lawyers who understand the intricacies of genetics and patent law is on the rise.

REFERENCES

Byrne, N., 1993, 'Patents for genes, other biological materials, and animals', *Trends in Biotechnology* 11:409-411.

Jorde, L. B., J. C. Carey, and R. L. White, 1997, *Medical Genetics*. St. Louis, Mo.: Mosby.

Paul, D. B., 1995, *Controlling Human Heredity*, 1865, Humanities Press.

Suzuki, D. T., 1989, *Genetics: The Clash Between the New Genetics and Human Values*, Cambridge, Mass.: Harvard University Press.

10

Genomic Valleys for India

On 26th June 2000 the whole world celebrated the announcement of the first draft of the human genome map, President Bill Clinton declared "this is the most wondrous map even produced by mankind". The Prime Minister of U.K. Tony Blair called the sequencing of the human genome "a breakthrough that opens the way for massive advancement in the treatment of cancer and hereditary diseases". Dr. Mike Dexter and Director of the welcome trust which pumped $120 million into the public project said, "A few months ago I compared the project to the invention of wheel. On reflection, it's more than that... this code is the essence of the mankind."

Mapping of genome sequence has revealed the biohoroscope of humankind, and at the sametime it has also advanced the future's biomarkets by many years. Looking at the giganticity of the task thousands of researchers in hundred of laboratories joined hands and a 'National Human Genome Research Institute' was formed near Washington DC. The Director of the Human Genome Project Dr. Frances Collins headed the Institute also. Human Genome Project was started in 1990 as a publicly funded sequencing effort with $300 million in the corpus.

ADVANTAGES

The positive impact of deciphering of human genome will be in the fields of health care and medicine. It will add a new branch of predictive medicine to the already existing branches of treatment curative and preventive medicines. Predictive medicines predictive branch will mainly concentrate on the identification the defective gene which causes a disease. So this will give us on opportunity to diagnose and predict the disease long before it actually occurred. This will set up a new curative model through gene therapy. It will be possible to identify the defective genes in a fetus to correct it before the birth of a child. It means that the defective gene would be replaced by the normal gene. Thus, uprooting the disease with the removal of causal agents. Now, it is possible to ensure that disease does not occur. Human genome information will be may crucial for finding out new curative methods and for developing predictive medicine. New genomic datas which will be available to scientific communities is definitely going to help not only in the fields of prevention of disease and discovery of novel drugs and therapeutics. It will also provide a window to the blueprint of life and evolution of human kind.

DISADVANTAGES

1. This information will be very advantageous for insurance companies and just by taking a drop of blood they will know that what kind of disease you may have in the future and that how far is that future. If the future disease is not very far they may not like to insure you.
2. Genomic information can be misused for creating discrimination within jobs.
3. It can be used by the state to impose some kind of controls.
4. This information is likely to used for creation of wealth.
5. People will start hunting for unusual traits.
6. Genomic value of those persons will increase who are not easily effected by diseases.
7. Celera Genomics has developed databases, access to which is very costly.
8. The structural genomics will give way to functional genomics with danger of patenting the sequences with known functions.

GENOME RESEARCH IN INDIA

Huge population of India can be on asset for us because this huge population of India has huge genetic diversity. This great genetic variation would help in utilising this great scientific breakthrough to remove human misery in our country because lot of genetic material is required for getting closes to develop genetic medicine.

We have the information in the form of biochemical script and now we can easily read it. So called jurk part of DNA might have some functions. Now, one should move from structural genomics to functional genomics. We couldn't make heavy investments in structural genomics but we can do better in functional genomics because we have the best analytical minds. We missed the 20th century Silicon Valley but with our vast scientific manpower we should concentrate more on 21st century genomic valleys.

India has a very strong infrastructure and expertise in this field. Initiatives in the field of screening of genetic diseases was first taken in 1990 and now. India is a position to genetically screen a large number of disorders. Fourteen genetic counselling clinics have been set up. A major programme on human genome diversity studies has been launched under Dr. Partha Majummdar in West Bengal and North Eastern States is which is first of its kind. His study has standardised many parameters. Development and standardisation of rotational screening, particularly of thalassemia, sickle cell anaemic, muscular dystrophies and neurological disorders already been undertaken. Indian scientists are already working to develop HIV diagnostic kits, candidate vaccine for malaria, candidate DNA vaccine for rabies, and a candidate multivalent vaccine for malaria, using genomic information. Department of Biotechnology has supported the programme of functional genomics at the centre of biotechnology studies. A team in the All India Institute of Medical Sciences headed by Prof. I.C. Verma, has done genetic screening of the population in Andman and Nicobar Islands and in the tribals of Bihar and

Orissa region from genetic counselling more than 60,000 families have already been benefited. Through genetic counselling already the incidence of Thalassemia in children has been reduced in collaboration with the All India Institute of Medical Sciences. Three different novel genetic loci which are believed to be associated with human meningoma, a form of brain cancer due to alteration in the human genome have been identified. Till now about 300 families have been benefited by its genetic counselling facility. Indian genetic counselling clinics will also be getting the advantage of the newly generated knowledge of Human Genome. The project on functional genomics is being implemented at the centre for biochemical technology where studies related to genomic instability, trinucleotide repeat expansion and neuro-degenerative disorder have opened up new vistas fast and efficient methods for isolation of DNA from 10 ml of blood samples and a PCR based analysis for disease detection have been standardised.

For development of genomic database for mirror sites have been created for functional genomics, which are collecting the genomic data from genomics point of view. Theses sites one at Jawaharlal Nehru University at New Delhi, Institute of microbial technology, at Chandigrah and Indian Institute of Chemical biology, Calcutta. Bioinformatics Network under the Jaivigyan Mission is accelerating research in areas of computational genomics and comparative genomics. These focused genomic segments will ultimately lead to novel DNA-based diagnostics and therapeutics. Gene therapy on oral cancer at Cancer Research Institute, Mumbai is progressing fast. A virus has been identified to be used as vector... and the experiment is being conducted. India has set up it's own Bioethics Committee which will consider any amendment in the universal declaration on human genome and human rights, make liaison with the International Bioethics Committees and consider programmes on Gene Therapy Research in the country.

Bibliography

Alberts, B., D. Bray, J. Lewis, M. Raff, K. Roberts, and J. D. Watson., 1994, *Molecular Biology of the Cell*, 3rd ed. New York: Garland.

Aravind L., Walker, D. R., Koonin, R. V., *Nucleic Acids. Res.* 27, 1223, 1999.

Avery, O. T., MacLeod, C. M. and McCarty, M., 'Studies on the Chemical Nature of the Substance Inducing Transformation of the Pneumococcal Types', *Journal of Experimental Medicine*, 79:137-158.

Battey, J., Jordan, E., Cox W. Dove., *Nature Genet*. 21, 73, 1999.

Beadle, G. W., and B. Ephrussi, 1937, Development of eye colours in Drosophila Diffusible substances and their interrelations. *Genetics* 22:76-86.

Beadle, G. W., and E. L. Tatum, 1941, Genetic control of biochemical reactions in *Neurospora: Proceedings of the National Academy of Sciences*, USA 27:499-506.

Bickmore, W. A., Sumner, A.T., *Trends Genet*, 5, 144, 1989.

Brown, K., 'The Human Genome Business Today', *Scientific American*, July, 50-55, 2000.

Byrne, N., 1993, 'Patents for genes, other biological materials, and animals', *Trends in Biotechnology* 11:409-411.

Campbell, J. L., 'Eukaryotic DNA Replication', *Annual Review of Biochemistry*, 55:733-771, 1986.

Collins, F., and S. M. Weissman, 1984, The molecular genetics of human haemoglobin. *Progress in Nucleic Acid Research and Molecular Biology* 31:315-462.

Cook-Deegan, R. M., *The Gene Wars: Science, Politics, and the Human Genome*, New York: Norton, 1994.

Craig, J. M., Bickmore, W. A., *Nature Genet* 23, 387, 1999.

Creighton, H.B., andMcClintock, ., 'A Correlation of Cytological and Genetical Crossing-over in Zea Mays', *Proceeding of National Academy of Sciences*, USA 17: 492-497, 1931.

Crick, E. H. C., 1962, 'The genetic code', *Scientific American* 207 (Oct): 66-74; Nirenberg.

Crick, F., *What Mad Pursuit: A Personal View of Scientific Discovery*. New York: Basic Books, 1988.

Crow, J. E., 'Mapping Functions', *Genetics* 125: 669-671, 1931.

Deloukas, P., *et al.*, *Science* 282, 744, 1998.

Edwards, J.H., *Ann. Hum. Genet*. 29, 77, 1965.

Eisen, J. A., Hanawalt, P.C., *Mutat. Res. DNA Repair* 459, 243, 2000.

Eisensmith, R. C., and S. L. C. Woo, 1995, 'Molecular genetics of phenylketonuria: From molecular anthropology to gene therapy', *Advances in Genetics* 32:199-271.

Ezzell, C., 'Beyond the Human Genome', *Scientific American*, July, 64-69, 2000.

Franklin, R. E., and R. G. Gosling, 'Molecular Configuration in Sodium Thymonucleate', *Nature*, 171:740-741, 1953.

Greider, C. W., and Blackburn, E. H., 'Telomeres, Telomerase, and Cancer', *Scientific American*, 267 (Feb): 92-97, 1996.

Griffith, F., 'The significance of pneumococcal types', *Journal of Hygiene*, 27:141-144, 1928.

Handin, R. L, S. E. Lux, and T. P. Stossel, (eds.), 1995, *Blood: Principles and Practice of Haematology*. Philadelphia: Lippincott.

Hayes, R., 'The politics of genetically engineered humans', *Exploratory Initiative on the New Human Genetic Technologies*, San Francisco, CA rhayes@publicmediacenter.org, 2000.

Hershey, A. D., and Chase, M., 'Independent Functions of Viral Protein and Nucleic Acid in Growth of Bacteriophage', *Journal of General Physiology*, 36:39-56, 1952.

Hofmann, K., *Cell. Mol. Life Sci*. 55, 1113, 1999.

Holtzman, N.A. and Marteau, T.M., 'Will Genetics Revolutionise Medicine?' *The New England Journal of Medicine*, July, 343, 2000.

Horvath, J.E., Schwartz, S., Eicher, E.E., *Genome Res*. 10, 839, 2000.

Howard, K., 'The bioinformatics goldrush'. *Scientific American*, July, 59-69, 2000.

International Human Genome Sequencing Consortium, *Nature* 409, 860, 2001.

Jackson, D. A., Iborra, F. J., Manders, E. M., Cook, P.R., *Mol. Biol*. 16, 54,1999.

Jones, N. L., CRS Long Report for Congress, RL30006, *Genetic Information: Legal Isuues Relating to Discrimination and Privacy*, 2 October 2000.

Jorde, L. B., J. C. Carey, and R. L. White, 1997, *Medical Genetics*. St. Louis, Mo.: Mosby.

Judson, H. E., *The Eighth Day of Creation*, New York: Simon & Schuster, 1979.

Kornberg and T. A. Baker, *DNA Replication*, 2nd ed. New York: Freeman, 1992.

Lindesley D., and G. Zimm., *The Genome of Drosophila melanogaster*, San Diego Calif.: Academic Press, 1992.

Marra M. A., *et al*., *Genome Res*. 7, 1072, 1997.

Merrick, W. C., 1992, Mechanism and regulation of eukaryotic protein synthesis. *Microbiological Reviews* 56:291-315.

Meselson, M., and Stahl, F. W., 'The Replication of DNA in *Escherichia coli*'. *Proceedings of the National Academy of Sciences*, USA 44:671-682, 1958.

Moore, J. A., 1993, *Science as a Way of Knowing: The Foundations of Modern Biology*, Cambridge, Mass: Harvard University Press.

Mullis, K. B., 'The Unusual Origin of the Polymerase Chain Reaction', *Scientific American* 262 (Apr): 56-65, 1990.

Nahon, J. L., *Crit. Rev. Neurobiol*. 8, 221, 1994.

National Center for Genome Resources Survey, *Attitudes Toward Genetic Testing*, March 4, 1998.

Nei, M., *Molecualr Evolutionary Genetics,* Columbia University Press, New York, 1987.

Olby, R. C., 1985, *Origins of Mendelism*, 2nd ed. Chicago: University of Chicago Press.

Orel, V., *Gregor Mendel: The First Geneticist*. Oxford: Oxford University Press, 1996.

Pandey, A. and Mann, M., *Nature* 405, 837, 2000.

Parker, J. 1989, Errors and alternatives in reading the universal genetic code. *Microbiological Reviews* 53:273-298.

Patthy, L., *Gene* 238, 103, 1999.

Paul, D. B., 1995, *Controlling Human Heredity*, 1865, Humanities Press.

Pavy, *N., et al.*, *Bioinformatics* 15, 887, 1999.

Peltonen, L., McKisick, V.A., *Science* 291, 1224, 2001.

Prota, G., 1992, *Melanins and Melanogenesis*. San Diego: Academic Press.

Ruddle F. H., and Kucheralapati, E.S., 'Hybrid Cells and Human Genes', *Scientific American* (Jul): 36-44, 1974.

Sakahira, H., Enari, M., Nagat, S., *Nature,* 391, 96, 1998.

Sambrook, J., Fritsch, E. E and Maniatis, T., *Molecular Cloning: A Laboratory Manual*, 2nd ed. Cold Spring Harbor, N.Y.: Cold Spring Harbor Laboratory Press, 1989.

Schuler, G. D., *Trends Biotechnol.* 16, 456, 1998.

Sturtevant A. H., 'The Linear Arrangement of Sic-linked Factors in Drosophila, as Shown by Their Mode of Association', *Journal of Experimental Zoology* 14: 43-59, 1913.

Sturtevant A. H., *A History of Genetics*. New York: Harper and Row, 1965.

Sturtevant, A. H., 1965, *A History of Genetics*. New York: Harper & Row.

Suzuki, D. T., 1989, *Genetics: The Clash Between the New Genetics and Human Values*, Cambridge, Mass: Harvard University Press.

Venter J. C., *et al.*, *Science* 291, 1304, 2001.

Wansink, D. G., *et al.*, *J. Cell Biol*. 122, 283, 1993.

Warurton, P.E., Willard, H.F., in *Human Genome Evolution*, Jackson, M.S., Strachan, T., Dover, G., (eds.), BIOS Scientific, Oxford, 1996.

Watson J. D., and Crick, F.H.C., *Nature*, 171, 737, 1953.

Williams, N.M., *et al.*, *Hum. Mol. Genet*. 8, 1729, 1999.

Woods, R. A., 1980, *Biochemical Genetics*, 2nd ed. London: Chapman and Hall.

Index